WARRIOR GODDESS

BOOK TWO OF THE MEI LING LEE TRILOGY

BRENDAN WILSON

Elite
PUBLICATIONS

Copyright © 2024 Fleet Publications, Inc.

www.brendanwilsonwrites.com

Book cover graphic illustration by Jeremy Ashmore
Cover and interior design by Krystal Harvey / Tiger Shark, Inc.

Library of Congress Control Number: 2 0 2 4 9 0 4 7 4 6

ISBN: 978-1-958037-19-5 (paperback)

ISBN: 978-1-958037-20-1 (hardcover)

KINDLE/EBOOK AVAILABLE

Published by

Elite Publications

Elite Publications
Greenville, NC 27858
Tel: 919-618-8075
info@elitepublications.org
www.elitepublications.org

PRINTED IN THE UNITED STATES OF AMERICA

For my lovely wife, Kay, gentle and kind; and known as Mad-Dog Kay by those who have fought against her. "You got this."

Human potential at its best is to transform a personal tragedy into triumph, to turn one's predicament into a human achievement.

-Victor Frankl, *Man's Search for Meaning*

ACKNOWLEDGEMENTS

To Cynthia Rothrock, whose life's work inspired a generation of women martial artists.

To Dennis Callanan, a great friend and fellow martial artist. I sent him an early draft just to ask for his thoughts. He came back with weeks' worth of line-by-line edits.

To Jeremy Ashmore who did an incredible job on the cover art.

To the Elite Publications Team who made this book possible:
- ❖ Jessica Phillips, Director of Marketing, who was more than patient with a fussy author.
- ❖ Krystal Harvey for a great job at graphic design, editing and formatting.
- ❖ Grandmaster Jessie Bowen for his indomitable spirit in getting this done.

To Stephen Wilson for his creative input at critical points in the novel, and for his knowledge of Mongolian history.

To Dr. Rebecca Wilson for her review of medical information.

To Claude Sturm for his advice on special operation tactics.

To Caryl Chambers and her family for allowing me to use the name of her late husband, Colonel Jay Chambers. Jay and I grew up together. When my parents died, his family took me in until I finished high school. He saved my life more than once. May there always be heroes like Jay.

-Brendan Wilson

GRANDMASTER CYNTHIA ROTHROCK
QUEEN OF MARTIAL ARTS CINEMA

Brendan Wilson's *Warrior Goddess* is a novel about a woman martial artist set in the far future. Though the setting is science fiction, the theme is timeless: a woman is thrown into a world of violence, confusion, and passion that men primarily dominate. The adventures of Captain Mei Ling Lee will ring true to any woman who has had to fight for her place in life or her chosen profession. She is faithful to her friends, her lovers, and her mission, yet just as flawed as any of her readers.

I first met Brendan at the Action Martial Arts Hall of Honors Mega Weekend, where he was awarded Grandmaster of the Year in front of an audience of 1,200 other martial artists. He is the right person to bring together the diverse strands of martial arts, military combat, and science fiction. Much of his vivid writing evolves from his experience and interests.

Brendan is a 25-year veteran of the U.S. military, a paratrooper and ranger, and a veteran of Iraq, Kosovo, Bosnia, Korea, Ukraine, Libya, and

Russia. Formerly a NATO diplomat, he served throughout Europe and the Middle East.

As a martial artist with over 50 years of experience, Brendan has done it all. He led a military competition team to both state and national honors and won the silver medal at the U.S. Open for Taekwondo forms.

The novel's depiction of martial arts is more than credible; it is real. The reader is immersed in it. You feel the sweat, sense the fear, and burn with the irrepressible fury of a warrior who will not surrender despite facing overwhelming odds. I know this feeling; I've been there.

For those readers who haven't read the first book in the series, The Achilles Battle Fleet, have no fear. Yes, you will want to read the first novel, but Warrior Goddess can be read as a stand-alone novel.

-Grandmaster Cynthia Rothrock

CHAPTER 1

A FLYING RESCUE

Alliance Marine Captain Mei Ling Lee was falling. She was encased in a small, transparent synthetic metal sphere, tightly packed in, her knees drawn up to her chin, which was accelerating toward a jungle planet in the Stygian star system.

Around her, spread out over several hundred meters, she could see her team: a dozen Marine commandos, each encased in their own spheres.

Twelve minutes earlier, they had performed a low orbital insertion from an Alliance Naval reconnaissance starship that had dropped out of light speed just long enough to spit them out before jumping back into faster-than-light, or FTL, travel.

The light synthetic material of the sphere protected them from the extreme heat of friction while they entered the atmosphere in free fall. Eventually, she knew the thin material would burn away, and they would be in true free fall.

Lee could feel the beginning of the buffeting as the ball engaged the thin air of the upper atmosphere. The shielding began to glow as it grew hotter, and she could see pieces of the material flake away upward in hot, bright sparks. Soon, the shell became opaque with flame, and she began to feel claustrophobic as if she was in a coffin on fire.

Then suddenly, a loud "pop" and the material disintegrated entirely. Relief flooded through her as she felt the rush of cold air. She immediately deployed the portion of her battle armor that transformed it into a wingsuit and assumed the standard skydiver's free-fall position: arms spread wide, legs apart. She felt herself stabilize and looked around to see that her team had done the same.

She breathed deeply and took a moment to acknowledge that she was happy. She loved free-falling; she loved being a Marine, but most of all, she loved the feel of impending combat.

"Status check; count off," she ordered to her teams on the internal net.

"Team One, secure," replied the sergeant in charge of that team.

"Team Two, secure," said another voice.

"Double-V formation, on me," she said.

In a well-practiced procedure, the two teams each formed a V formation behind her, one behind the other.

Lee used her terrain mapping radar display, embedded in her helmet and heads-up display in her visor, to pick out the key terrain features on the ground that she needed for the insertion. Quickly, she identified the valley they would follow to their destination, a camp of smugglers that had been holding enslaved people to sell on the black market. In the two years since the surprise attack on the Alliance and the civil war that followed, the outlying state systems had fallen into lawlessness.

She realized she had several minutes of free-fall before she would need to lead her team through the treacherous mountainous terrain toward their objective. Her mind recalled the past 18 months since her transfer from the Navy to the Marine Corps, her demotion from Naval Lieutenant Commander to Marine Corps Captain, her trial and acquittal at court-martial, and her new career and life as a Marine Officer.

While Lee was still in the Navy, she had led a Marine Corps special operations task force. It was the events of that battle that led to her court martial. The Navy had tried to arrest her at the spaceport when she returned to Earth. When that failed, they sent assassins after her. But the Chinese government and the Marine Corps had intervened.

Her transition to the Marine Corps had been a rough time, harder by far than she could have imagined. She said from the beginning that she wanted to command a Marine company. And for her to do that, she had to qualify via a number of highly demanding courses.

There had been the Ranger Course, 58 days of training as a paratrooper and ground warrior. She learned the intricacies of infiltration, patrolling, reconnaissance, and raids. The candidates were permitted little sleep or food and were randomly inserted into leadership roles for evaluation during training missions.

Then came sniper school, where the Corps disabused her of the notion that she could shoot accurately. During the battle of Alpha 51, she had efficiently and accurately used her weapon against the enemy. Lee thought she was a reasonably good marksman, but combat sniping was its own unique world of discipline, patience, and deadly precision.

Finally, she attended the Marine Corps advanced command school with six months of intense leadership training. She had learned tactics, maintenance, logistics, personnel, and report writing. But most of all, personal leadership. Marine officers were expected to lead from the front. The Marines operated in small groups, rarely larger than battalion size, but mostly platoon and company-sized organizations: 40 to 100 Marines. Combat tended to be intense, and officers bore the same burden of risk and stress as the Marines they led.

When she graduated from the advanced command school, she initially got orders to report to the office of the Marine Commandant. She was devastated. It sounded like a desk job fetching coffee for colonels and generals.

But it turned out otherwise. When she reported, the Marine Corps Commandant, General Mosi Motubu had greeted her warmly. Motubu had personally intervened when the Navy wanted her head after the battle of Alpha 51. Those who were behind the conspiracy wanted her tried as a distraction from their own guilt. Motubu arranged for her transfer to the Marine Corps so she could be tried by an impartial judge. Once acquitted, neither the Navy nor any other court could not re-try her because of the rule prohibiting double jeopardy.

Motubu smiled and said in his melodious East African accent, "Captain Lee, it is good to see you. I have heard good reports on your training. Are you ready to go to work?"

Lee stood at attention and responded, "Yes, sir. But I had hoped to lead a combat unit."

He laughed and said, "What? Did you think I would put you on the staff? Here?" He shook his head and said, "No. You are here only so I can brief you and send you out. Please come. Coffee?"

Lee said yes, and she was surprised that he went to a side table and poured two steaming cups from a metal pot and a warming plate. No servants for this man. She recalled that the East Africans would not have personal servants or anything that smelled of it.

Lee knew that Motubu was descended from the East African tribes that had survived the turbulent times after the great collapse of the world order in the 22nd century. The tribes had survived through extreme toughness and, yes, violence. When a rival tribe tried to enslave members of Motubu's tribe, the reprisal had been unforgettable. Even now, two centuries after the Alliance had reestablished order, the tales of those tribes during the troubled times were only hinted at, spoken of in whispers.

At first appearance, Motubu was a quiet, self-contained, and friendly person. He was less than average height but thickly muscled and broad-shouldered. Once, during a terrorist attack on the Chinese Embassy in Colorado, Lee had watched him snap an attacker's neck like it was a

4

twig. And the next moment, he was his calm, polite self, apologizing to the hostess for the mess.

"Sit and let's talk," he said. She sat in a comfortable yet functional chair, and he sat opposite her in a similar one.

"First, tell me of your training."

Lee briefed him about her experiences in the courses she had been through, highlighting her challenges and what she had learned in each.

"Yes, I must confess I already knew. I have been tracking your progress. When you spoke of it just now, you left out that you were at or near the top of your class in each course."

Lee blushed slightly. Her upbringing in a traditional Chinese household had strictly forbidden any behavior that could be interpreted as self-promotion or, God forbid, boasting.

She stammered and said, "To be fair, I had the support of many of my colleagues, without whom I could not have succeeded or even graduated."

Motubu responded with a serious look, "Yes, I know this too. Your ability to work in a team speaks well of you. If you had been a *spotlight ranger*, you would not be here now." A spotlight ranger was a derogatory term for a student at ranger school who was sluggish and unhelpful to others but perked up when in a leadership position or when the instructors were nearby.

Motubu paused a long moment before speaking. It was a technique she had seen her former boss, Admiral Jay Chambers, use when he wanted someone's full attention.

"The situation is very bad; it has gotten worse since your court martial." He looked hard into her eyes and then continued. "The separatists have solidified their positions in a number of star systems. The fighting has been sporadic and chaotic. Regional governments have broken down, and with it the orderly functioning of society has also suffered. Trade was disrupted, and that, combined with a breakdown of the rule of law, has led to a startling increase in organized crime."

Lee asked, "What is the role of the Marine Corps in all this?"

"That is a good question," said the general. "I'm afraid the answer is unsatisfactory. On the one hand, we are uniquely qualified to conduct counter-terrorist operations, and this can be adapted to anti-crime. We can interdict criminal operations, but without some civil governance, the crime rings just naturally re-emerge with new leadership."

"So, what is the solution?" said Lee.

"For the moment, we must continue as we are. We have no choice," he said. "But we are working — slowly — on a strategy of finding and supporting civil leadership to fill the vacuum which we ourselves are helping to create." He shook his head and said, "And this is a big task, something we are *not* suited for."

"Sir, if I may ask, why am I here?"

Motubu laughed and said, "Now that is the Mei Ling Lee I remember and admire. Very direct."

Continuing in a serious voice, he said, "At some point, I will need you, and others like you, to step up to a more diplomatic approach. But for now, you will need to learn your trade and develop the gravitas that only comes through leadership in the field."

The general stopped talking, looked away as if he were deciding something, and then said, "I am sending you out to command a recon company in the Stygian system. It is a brutal assignment. The company has lost all of its officers twice over to combat in the past year. And the tasks it must take on are often very hazardous and without much, or any, support. I trust that you will lead well, and I hope that you will survive. I will need you later."

Six hours later, Lee was on a starship traveling FTL bound for the Stygian system.

CHAPTER 2

DEATH FROM ABOVE

A beeping in her headset brought Lee back to her situation. "Coming up on Checkpoint Alpha," said her second in command, Gunnery Sergeant Ian MacGregor.

MacGregor was an interesting guy, thought Lee. He was 45 years old, middling height, face weathered and lined. He was unusually quiet for a senior NCO. He got things done readily and without fuss. And with Lee, he was completely and always correct. If he had any reservations about having a female company commander half his age, he never showed it. The Marines in the company obviously had confidence in him. Before Lee had arrived, he had been the acting company commander for several months because all of the commissioned officers had been killed in combat.

Reviewing the records after her arrival, Lee could see he had done a good job as a company commander. With the advent of war and the creeping casualties in the Marine Corps, many senior NCOs had received commissions or were encouraged to go to officer candidate school. Lee saw from his personnel file that MacGregor had turned down both of those options.

"File formation," said Lee. "Snipers left and right."

Lee flew into the narrow valley that led to their objective, taking point as was expected for a Marine officer who was to lead from the front. The reconnaissance from space had shown that the smugglers would have lookouts posted along the valley. Lee knew that her snipers would be ready if they came across any along the route.

The Marine snipers worked in teams of two, with a spotter and shooter. At their current relative speed of 100 kilometers per hour, the Marine shooter would have only a moment to acquire and take out a lookout.

"Target left, ten o'clock," she heard over the tactical net, then a muffled shot, then another voice, "Target down."

This was something Lee had learned from her time at the sniper school, but it still amazed her. Marine snipers — her snipers — had just hit a target while flying in a wingsuit at 100 kilometers per hour. The shooter had at most two seconds to acquire and shoot the lookout. The mechanics of the shot were outrageously complicated. And yet, she knew they would not miss.

When she first saw this demonstrated at the sniper school, she asked the instructor how such a shot was even possible. The sergeant had responded, "A fully trained Marine sniper could no more miss that shot than you could fail to catch a ball thrown directly at you."

Although she passed the course, she had nowhere near the skill of the Marines on her team. Their skill was supernatural and spooky to her. She was glad they were with her and not against her, and she found herself selfishly hoping they would not be killed.

Twice more, she heard the radio exchange of spotter and shooter, each time, a confirmed kill.

"Objective in 30 seconds," said the laconic MacGregor.

"Roger," Lee responded. She switched to the company net and said, "First platoon, status?"

Immediately she heard, "First platoon engaging now."

Her first platoon, a group of 20 Marines led by a staff sergeant, had been covertly inserted on the ground two days ago. The plan called for them to attack the far side of the base's perimeter at dawn to draw out the defenders to that side while Lee came in with her team behind the defenders.

Lee could hear the gunfire and explosions the Marines were making over the radio transmission.

"V formation, attack profile," said Lee. In response, the dozen marines formed a single wedge shape with her at the point. The plan was to strike the defenders and shock them as a single unit.

CHAPTER 3

SORRY, DOES IT HURT?

As Lee and the team rounded the last outcropping of the valley, the garrison came into view. Lee saw what looked like a smallish fortification made up of a few buildings with a landing pad for aircraft, surrounded by a wall about three meters high. She could see figures running about in what looked like confusion on the open ground and another twenty or so on the wall's rampart firing into the nearby wood line.

"Weapons up," said MacGregor in his steady voice. The Marines raised their weapons to the firing position. "Targets." This was the order to acquire their targets. "Fire!"

As one being, the Marines fired. Lee saw the impact immediately. A dozen of the smugglers dropped. By this time, Lee's assault force was almost on top of the fortress. At her lead, they slowed and set down in the center of the open area. They quickly formed a circular defensive perimeter and continued firing.

"Teams forward!" She yelled. Each team moved out toward the clumps of surviving defenders. There was almost no return fire. Clearly, these were smugglers, not warriors.

Those on the wall who had been firing into the tree line toward the Marine platoon were the most vulnerable; they were exposed without any cover. Some of them tried to turn toward the assault force and return fire. They didn't last long.

Quickly, Lee gave the order that would stabilize the situation. "Alpha Team, secure the buildings. Bravo on perimeter. XO, have first platoon sweep the outer perimeter. Medical team on me."

Lee could see that her own Marines didn't have any casualties, but they would offer treatment to any of the smugglers who had survived and to any of the captives they had been holding as enslaved people to be sold.

Just then, there was a disturbance near one of the structures. She saw her Marines drop into combat stances and take aim at something that she couldn't see near one of the doors to a windowless structure.

She heard a Marine command, "Drop the weapon, hands in the air, do it now!"

This command surprised her because, under the current rules of engagement, the Marines were not obliged to accept a surrender. Any smuggler who was still armed could, and should, be shot without warning.

As she moved toward the disturbance, she saw why that command was ordered. An overweight, disheveled, middle-aged man was holding a girl about 13 years old. He had a knife to her throat. The blade gleamed in the early dawn light, and Lee could see blood was already trickling down the girl's blouse. The child was filthy, dressed in rags, and she looked terrified.

The man said, "Back off, or I will cut this little bitch's head off."

Lee came forward and stood about ten feet in front of the man. She looked at him with a cold face, the look that showed nothing, neither anger nor fear. It was the look her ancestors, the Mongol warriors, had used going into combat. She said nothing.

The man sneered at her and said, "Oh, please don't tell me the Marines let Chinese bitches lead them."

Lee said nothing.

The man said, "Listen, all I need is that you let me go. I'll take one of the helos. And I've got to take this lovely with me." He pulled the girl's hair back for emphasis. "Gotten used to her, I have. Keeping my bed warm at night. You won't miss her. You can have the rest." He nodded toward the building he had just left, and Lee saw several dozen frightened girls and women cautiously leaving the building, shielding their eyes against the dawn sunlight. They too, were filthy and dressed in rags.

The man looked at Lee and said, "No? Well, then off with your head, lass." He pressed the knife deeper, causing the blood to flow.

"Hold on there," said Lee. She raised both her hands, palms outward, in a pacifying gesture. Then, she turned her head and said in a loud voice, "Sergeant MacGregor!" The man turned his head to look where she had indicated.

"Ma'am," replied MacGregor.

Lee then snapped her right hand downward, causing a narrow spike about four inches long to fly out of her sleeve toward the man. It struck him squarely in his left eye, penetrating two inches into the socket.

The man screamed, threw the girl to the ground, and raised his hands to his face. Blood and viscous fluid gushed from his eye.

Lee approached him and said in a gentle tone, "Oh, my. Does it hurt? It must. Let me see if I can help." Then she quickly tapped the base of the knife handle with her palm, driving the blade another inch into the socket.

The man screamed and began to beg, "Please, no. I'm sorry, you can have the bitch. She wasn't that good anyway."

"I'm so sorry, sir," said Lee. "Let me fix it." Then she drove her palm hard, pushing the knife deeply into the man's skull. He fell backward onto the ground and began convulsing.

For a long moment, there was silence. Then the girls as a group came out quietly and surrounded the twitching man. First one, and then the others, methodically began kicking and stomping the downed man, their voices rising in cries of fury.

CHAPTER 4

A NEW MISSION

An hour later, the base was secure and two Naval shuttles had landed. One with a full medical crew to take charge of the girls and young women who had been held as slaves, and the second to return the Marines to the troop transport that served as their base.

"Ma'am?" said Gunnery Sergeant MacGregor.

"Yes, Gunny," replied Lee.

"You have an encrypted, eyes-only link from CMC actual."

CMC was an abbreviation for Commandant of the Marine Corps. The actual message indicated the caller was the general himself, not somebody from the Commandant's staff. If Sergeant MacGregor was surprised that the commandant of the Marine Corps would hail a lowly Marine captain, he didn't show it.

Lee looked at her hand-held communicator and said, "I've got it." She walked away from the group of Marines near her and then activated the link. What showed was a hologram about four inches tall displaying the head and torso of General Motubu.

"Captain Lee, I am gratified you have survived another mission."

"Thank you, sir. And I trust you are well," she said. She felt awkward exchanging such pleasantries. Lee was still feeling the adrenaline rush of the drop from space and the combat, and she was struggling to make the transition to this call and its likely import.

"Yes," said Motubu. "Unfortunately, I need to remove you from your current duties because I have other pressing requirements." *That was Motubu*, thought Lee. Always to the point. Her heart sank at what she knew was coming.

"You will need to hand over command to your XO. I'm sending a shuttle for you now." Lee actually saw a third shuttle moving up the valley toward her position. *Not wasting any time*, thought Lee.

"What's the mission?" said Lee, deadpan. Her full ambition for the past few years had been to command a Marine company. She had finally achieved that and was doing a reasonably good job of it. It was all she really wanted out of life. And now that dream was ending. She was beginning to get angry, and that was not a good thing, she knew.

"I told you when we last met that I would need you, and others like you, for the next phase of our engagement," said Motubu. "That time has come. On the shuttle there is a naval intelligence officer who has the mission parameters. He will brief you and accompany you as far as possible."

"I don't want to give up command," she said simply.

Motubu was silent and looked expressionless at her for several long seconds. Then he said, "That is understandable."

He paused again and then said, "The mission you will be assigned to is important and, I regret to say, more challenging and hazardous than the command you now hold. You have been in command six months..."

"Yes, and a command tour is supposed to be two to three years," Lee interrupted.

The general looked at her steadily. It was a rare thing for a captain to speak so abrasively to a four-star general.

He continued without acknowledging her interruption, "In that time, one-third of all company-grade officers in your sector have been killed in combat. I am running out of marine officers, and the replacements from direct commissioning and the officer candidate school are not enough to make up these losses."

"I'm the only officer in this company," Lee said. She knew this was a losing battle, and she knew she was whining in a way she should never do.

"Your XO, Gunnery Sergeant MacGregor, has in the past, and will again, ably command your company. He can command a company, maybe a battalion; he's very solid. But I cannot send *him* on the mission now before you. I'm sorry, Mei Ling; you will simply need to do as you are told. This is what we do."

Lee recalled the phrase, '*This is what we do*,' was something Admiral Chambers had said when he took command of the Achilles Flotilla after all the other flag officers had been killed when the war started.

She straightened a bit and said, "Yes, sir. I'm ready."

"I knew you would be," said the general. "Please call Gunnery Sergeant MacGregor over so we can have a word."

Lee turned, saw MacGregor, and waved him over to her. When he arrived, MacGregor immediately saw a holographic image of the general and stood at attention.

"As you were, Gunny," said Motubu. "I need to pull Captain Lee out of the line for an important mission. You will take command of the 75th Reconnaissance Company."

"Yes, sir," said MacGregor with a blank expression.

"Gunny, I know your record, and I know you have commanded this company twice before when no officers were available."

"Yes, sir," said MacGregor. "If you don't mind, sir, I was hoping to hold onto Captain Lee for a while longer." And then Lee saw him do something she had not seen him do in the months she had been in command: he smiled.

"Gunny, I can't keep rotating officers through your command," said Motubu. "It's not fair to the unit, and it's not fair to you as the senior NCO in the unit."

MacGregor remained silent. He understood this was a point in the conversation that didn't require a response.

"I know you have turned down both OCS and a direct commission. I respect that, but I'm afraid the situation calls for a sacrifice on your part. Do you understand, Gunny?"

MacGregor sighed, paused a moment, and replied, "I do, sir."

"Right," said the general. "As of now, Gunnery Sergeant Ian MacGregor, you are promoted to Captain of Marines. Captain Lee is relieved, and you will assume command of the 75th Reconnaissance Company. The orders appointing you to your new rank and command have been transmitted to you. Do you have any questions, Captain MacGregor?"

"A request, sir," said MacGregor. "Please don't send me any lieutenants. I've got the platoons commanded by staff sergeants, and that's just how I want it."

The general smiled and said, "Well, you'll take whomever the Corps decides to send. But there is zero risk of getting any lieutenants anytime soon. We just don't have any available."

"Aye, Aye, sir," said MacGregor.

CHAPTER 5

THE WIDOW MAKER

Lee had a very short handover of command to now *Captain* MacGregor. Lee was surprised at how quickly MacGregor seemed to adjust to his new rank of captain. As an NCO, he had been perfectly proper in his dealings with her in every way. And now, as a Captain - her equal - he acted perfectly at ease. He didn't accidentally address her as "ma'am," nor did he show in his deportment any deference, but simply the easy mannerism of one equal to another. Lee was impressed. MacGregor had turned down a promotion to officer rank until it was made clear by General Motubu that his duty required him to accept the new rank. And now that he had that rank, he acted as if it was the most natural thing in the world.

Even though she knew she was expected to report to the shuttle to meet with the naval intelligence officer, she took the extra time to ask him how he felt about the promotion. "I'm a Marine," he said. "That's all. It doesn't matter what rank or job. I'll simply do my best to get the job done. This company doesn't need a reluctant captain; it just needs a competent captain. I turned down a promotion because that was my preference. I was comfortable as a sergeant, and we had officers, so I didn't see the need."

"And now?" Lee asked.

"Now, all the officers are dead, and you are being taken away," he gave a rare smile. "Do you know what the Marines call you?"

Lee shook her head.

"The *widow maker*," said MacGregor. "They like it that you are aggressive. That you go for the jugular and that you take the point and go first every time." He shook his head and said, "Since you took over, this is the happiest they've been in a long time, probably ever. They like what we're doing, and they look forward to operations. I knew if you were killed or moved out, I would step up as an officer if it were offered. I couldn't risk that we would get another commanding officer of, let's just say, lesser caliber."

Lee was stunned at the frank assessment. And she was pleased. She desperately tried to hide the blush that she felt warming her face. She had been awarded the silver star for her actions on Alpha 51, but MacGregor's words meant more to her than any medal. She felt again the frustration and sorrow of leaving command. She had finally found her place in the universe, and it was being taken away from her too soon.

But she felt the wisdom of his words. Like him, she was a Marine, *that's all.* Marines didn't get to pick their desired job; they just did the best they could at any assignment they were given.

And now, she was about to find out what her next job would be.

Lee turned toward the naval shuttle that had been sent to collect her. It had set down on the landing pad inside the perimeter that the Marines had already established, so in theory it was secured by Lee's — now MacGregor's — Marines. But the shuttle had its own security personnel, black-clad naval provost guards. They had taken up what Lee judged to be a reasonable defensive perimeter around the shuttle. *Understandable*, Lee thought. A few minutes ago, the area had been the scene of a firefight.

Lee felt an unease that was rare to her. The last time she had any significant dealing with the Navy, there was an unpleasantness. A naval commander with a squad of military police had tried to arrest her at the Denver spaceport when she returned to Earth after two years away in space. They intended to charge her with treason for the torture and murder of civilians, crimes that carried the death penalty. Had it not been for the intervention of the Chinese embassy which granted her diplomatic immunity, Lee's life would have taken a different turn.

But there was nothing to do now but do as she was told. She walked toward the shuttle.

CHAPTER 6

WAITING FOR LEE

Naval Lieutenant Andy Danner sat in a shuttle on a planet in the Stygian system, awaiting Marine Captain Mei Ling Lee. Danner had been tasked with controlling and supporting her upcoming mission.

Danner knew Lee. She had recruited him to be part of the intelligence section on the Achilles Battle Fleet in the aftermath of the attack of 7 December 2541. Danner had been a seaman apprentice, the second lowest rank in the Navy. He was part of a convoy withdrawing personnel from Achilles Nine, a remote outpost of the Alliance. The convoy had been ambushed and the command vessel had been hit by a suicide bomber carrying a nuclear payload. The commanding admiral and almost all of the command staff had been killed.

Rear Admiral Jay Chambers had assumed command. After Chambers had fought off the ambush, he rebuilt the convoy into a battle fleet. He had appointed Lee as his chief of staff and tasked her, among other things, with creating an intelligence section. She had picked Danner partly because of his education. He had a dual degree in engineering and in ancient history before joining the Navy. In a short time, he fell into the leadership role for the intel section. It was awkward because he was the lowest-ranking person in the section.

That awkwardness was solved when Chambers had ordered Lee to set up an officer training program on board, and Danner had been appointed to the rank of Midshipman, which allowed him to have the appropriate rank to lead the section. Later, prior to the battle of Alpha 51, Danner had been given a temporary promotion to ensign and had been tasked with heading up the combined operations/intelligence section.

During the battle, Chambers had ordered the crew to abandon ship, and Danner had supervised the evacuation and kept the hundreds of evac pods together and safe.

Following the battle and subsequent return to Earth, Danner attended the Naval Academy, reverting to his rank of midshipmen. Because he already held a degree and because he had experience in combat as an officer, he had been allowed to enter the academy as an upperclassman.

Danner had graduated quickly and been assigned as an intelligence officer. He had attended a six-month course and, upon graduation, he was informed that his date of rank would be calculated from the date of his appointment as an ensign by Admiral Chambers, not from his graduation from the naval academy. Because of the wartime expanding line, he was promoted to naval lieutenant, the equivalent of a Marine captain — Lee's equal.

And now he had been singled out for this assignment. He wasn't formally trained as a controller for covert operations. Handlers were typically people who had years of undercover experience themselves. Danner had none. His specialty was supposed to be strategic intelligence. But the circumstances were specific to Lee and, to some extent, to him.

With him on this mission was a naval doctor, Lieutenant Laura Zakany. Danner inwardly sighed at the thought of her. She was difficult and temperamental. Zakany had also served on the Achilles Battle Fleet. During the original evacuation from Achilles Nine, a remote Alliance outpost, she had been a civilian passenger on what, at the time, was only a convoy. Because she was a nurse practitioner, she had offered to support the onboard medical station during the passage back to Earth.

On the day of the attack, Zakany had been away from the medical ship and had thus survived when that vessel had been destroyed. As one of only three surviving medical personnel, she had been inducted into the Navy as a medical officer, third class, the equivalent rank of a Navy lieutenant.

Danner had to admit to himself that Zakany was a superb doctor. As a nurse practitioner who had been trained in her profession in the slums of Budapest, she was ideally suited to recruit and train new personnel in the various medical specialties required for an emergency room. Although she did not then have a medical degree, she was addressed as a doctor while on board.

But Zakany was more than a good medical practitioner. She was smart; no, she was brilliant in other ways. In the run-up to the battle of Alpha 51, Zakany had worked with captured clones to uncover their programming. She had developed a combination of verbal and visual cues that could be used to disable them.

During the battle, her mechanism had been used to disable the clone that was masquerading as the commanding admiral of the 5th Fleet. Then, once the battle was underway, she and Danner had together developed a computer virus that had infiltrated the enemy communication system that had allowed the Achilles Battle Fleet to defeat a superior force.

Alas, despite their good work together, there somehow was a standing animosity between them. At the time of the battle, he was an ensign while she was a newly minted lieutenant. She was aware and sensitive to the fact that she outranked him. She bristled at every order he gave in his role as the chief of operations and intelligence.

After the return to Earth, she had enrolled in the medical college in Perth as a student to complete her medical degree. Like him, she had advanced standing because of her education and experience and had graduated quickly.

But something had changed. Danner knew she had been married to warrant officer Matthias Nemeth before the battle. Nemeth and Admiral Chambers had gone missing after the battle. The Navy had declared them to be dead, but Zakany had never accepted it. She insisted he was still alive, and that was part of the reason she had volunteered for this mission.

But there were other changes. Danner knew she had had a child while in medical school. That child should be about two years old now. Danner wondered at that, and also that Zakany seemed withdrawn, distant. She was thinner than he remembered, and in the weeks since they had started this mission, he had rarely seen her smile.

Danner heard the naval provost guard come to attention, and the petty officer said, "Good morning, ma'am." He knew Lee had arrived, and he was about to see her for the first time in two years.

CHAPTER 7

ZAKANY

Naval Lieutenant Doctor Laura Zakany sat quietly in the shuttle across from Lieutenant Andy Danner. The last few weeks had been difficult. No, they had been horrible. When the naval intelligence agents had come to her in Perth, where she was on the teaching staff at the medical center, her life changed in a moment.

Not that life had been perfect. Her husband, Matthias, had been missing since the battle at the Moon Base Alpha 51. The shuttle that he piloted along with its other occupant, Rear Admiral Jay Chambers, had disappeared in a confusing engagement, which resulted in the loss of the command ship, the *Centurion*, and its shuttle. The *Centurion* had crashed into the Moon. A naval investigation found no remains of the Admiral or Warrant Officer Nemeth among the debris. Despite that, the report had concluded that Chambers and Nemeth had likely died in the crash, and they had both been declared dead.

Laura Zakany had never accepted that Matthias had died. In their final moment together, when she had to evacuate the *Centurion*, and it was clear he would remain with the Admiral, he promised to return to her. She believed him. The Admiral had married them the day before. But at the time, neither had known that she was pregnant. Laura had been sure he was out fighting the clones with the Admiral in some scheme of theirs.

The investigation had produced no DNA and no shuttle debris. How had the Navy been so sure they were dead? When a priest and an officer had come to notify her of her husband's death, she had thrown them out of her quarters and called them idiots.

But the bright light of her life was her child, *their* child, Mathias, Jr. He was intelligent and wonderful. Her joy was now with her Aunt Anna who was staying in her quarters with Mathias in Perth. And now she was light years away from her son, on this mission with the young upstart, Danner.

Danner had always irritated her. When she was a newly promoted naval Lieutenant on the Achilles Battle Fleet, and Danner was only a midshipman and later an ensign, he always seemed to her to be preening and seeking attention. It infuriated her that some of her work for the Admiral had to go through him and his team of so-called "smart guys."

He had changed in the two years since she had seen him last, and she had to grudgingly admit to herself that he had matured somewhat. He had filled out a bit, shoulders broadened, the acne gone, and now sporting a military haircut. Now, at least, he looked the part of a young naval officer.

On his uniform, he wore the naval commendation medal, an award she knew he had earned during the evacuation of the *Centurion* and subsequent leadership of the hundreds of people in escape pods during the battle. He had guided them to a point of safety in orbit around the gas giant, Lilith. In the chaos of the battle, it was no small feat.

What he did not know was that she had been the one who recommended him for the medal. Her dislike of him at the time did not prevent her from recognizing his leadership in a time of crisis. *Fair is fair*, she thought.

Zakany herself had been awarded a commendation for her development of the key to controlling, or at least disabling, the clones and for her part in developing a computer virus that infected all communications systems in the vessels where the clones were serving.

And now she waited for her best, and perhaps only, friend, Mei Ling Lee. When she first met Lee, Laura did not like her. Mei Ling seemed like this tiny Asian woman with long dark hair and a confident demeanor. She thought Lee put on airs and, if she were honest, she was just a tad jealous of her at the time.

But Lee had proved a good friend and confidant. Zakany and Lee were two women in a wartime environment that was dominated by men. They had become close, shared their hopes and fears, and gone into battle together. When Mei Ling had been wounded during the battle of Alpha 51, Laura realized what they meant to each other. She had called Mei Ling, *sister of my heart*, and she meant it.

Laura and Mei Ling had met only once since the return to Earth after the battle. That was in the officers' club in Perth. Zakany was heavily pregnant at the time with Mathias, Jr., and Lee was on a short break from her training with the Marines.

Laura did not know the whole story behind Lee's transfer to the Marine Corps and her demotion from naval lieutenant commander to Marine captain. Mei Ling didn't explain, but Laura had heard rumors that the Navy had tried to arrest Lee at the spaceport in Denver when Mei Ling returned to Earth. The Chinese government had intervened to protect her, granting her temporary diplomatic protection.

Afterward, at the Chinese consulate in Boulder, some sort of kidnapping attempt had taken place, thwarted by a Marine detachment detailed to protect the compound. To protect her from the wrath of the Navy, General Mosi Motubu, the commandant of the Marine Corps, had arranged for Lee to transfer to the Marine Corps where she could be tried by an impartial Marine Corps judge rather than a corrupt Naval tribunal. As part of that process, Lee had been offered an equivalent rank of Marine major but had refused on the grounds that she wasn't qualified to be a major. Instead, she had entered as a captain in the Marines.

That was two years ago. Now she waited for Lee in this shuttle, a galaxy away from her son. She heard the naval guard come to attention for Lee and stood to await her friend.

CHAPTER 8

A REUNION

Lee walked up the ramp of the shuttle. She was still in her light battle armor, covered in dust and a smattering of blood. Both Danner and Zakany stood in the tight space of the cargo bay. The three looked at each other. For a long moment, no one spoke.

Then Laura rushed forward and hugged Mei Ling, pulling her tight, silently crying. Though Mei Ling was not a hugger, she hugged back. Danner stood mute and looked away, letting them have their moment.

Finally, after a long moment, Lee pushed Laura away, holding her shoulders and looked her in the eye.

"What the hell are you doing out here?" asked Lee with an amazed smile.

"Mei Ling," replied Zakany with a smile. "I have missed you, too. Shall we let the lord Danner explain?"

Both women turned toward Danner.

Lee saw her old protégé and smiled approvingly. She noted the new rank and Danner's military bearing. Inwardly, she felt satisfaction for his success. She had picked him for the intel section at a time when he was an awkward, scruffy-looking seaman apprentice.

But the brilliance was always there. She had intentionally sent him in to interview for the position with a Navy captain without telling Danner the reason for the interview. She felt instinctively that Danner would respond better if he were allowed to answer questions without preparation. He had done well, even giving a sort of education to the captain on the difference between strategic and tactical objectives, and the need to formulate priorities based on higher-level political guidance.

But Danner was more than just smart. He was steady under pressure. He had been with her in the command center when the clones had broken through the provost guard and attacked. While Lee led half the team repelling boarders, Danner had calmly stayed with the other half of the section controlling the external battle.

That engagement had been close combat, and Lee had lost half her team, killed or wounded in a matter of minutes. Only the arrival of a Marine reaction team had saved them.

And when she commanded the special operations force that conducted the raid on Alpha 51, his was the voice from the operations center back on the *Centurion* calming her as she drifted in space for a terrifying few minutes as the team inserted from a passing asteroid.

Despite that history, there seemed to be no real connection between them. Why was that? She couldn't say.

"So, Lieutenant Danner," said Lee. "You're the Naval intel officer come to collect me." It was a statement, not a question.

CHAPTER 9

TIME FOR A NAP

Danner ordered the shuttle to depart quickly. They strapped into seats facing each other in the confined space of the shuttle.

"Isn't your security detail coming?" asked Lee as the door to the shuttle closed. She noted that only a pilot and co-pilot remained.

"No," said Danner. "They aren't cleared for this part of the mission. Another shuttle will come for them."

As the shuttle lifted off, Lee saw the compound through the window. Her Marines — no, check that, they weren't *her* Marines — were busy securing the site, giving medical aid to the newly-freed captives, and collecting the bodies of the slavers for burial. In a moment, the shuttle was away and the blue-green planet rapidly diminished.

And then, as the shuttle turned, the planet was gone from her view. She felt a pang of sorrow and loss having to leave them. She comforted herself that at least MacGregor was a competent commander. She would miss him, too, the steady, unemotive way he dealt with her and with everything.

She recalled an event early in her command tour. The company was in a jungle at night setting up a defensive perimeter. She was in the command post going over the orders for the next day's operations when she remembered that she had forgotten to issue orders to send out a security patrol. She moved in the dark until she found Gunnery Sergeant MacGregor huddled with the platoon commanders. She listened, unseen in the dark, and realized he was already issuing the order for the patrol.

The next day she asked him about it. He responded, "Ma'am, you do not need to issue every order. Security patrol is standard operating procedure when in a defense perimeter. Your job is to look forward; mine is to take care of current business."

And that's what they did. They worked well together, and she was grateful he had been her deputy in her first command.

Suddenly, Lee felt exhausted. She hadn't slept in about 30 hours, and the stress of the drop from orbit and the combat raid seemed to hit her all at once.

She turned to Danner and said, "I'm going to sleep. Wake me in an hour." Then she closed her eyes and, in a moment, was fast asleep. It was a technique she had learned in Marine training. She had seen the Marines use it before the battle of Alpha 51. They were on an asteroid spinning through space, getting ready to jump off and make a moo-suit infiltration into combat on the Moon. The sergeant major had told them they had a few hours, and they all simply went to sleep. Lee learned this trick in the extremely demanding ranger course. For months on end, the students were not afforded any actual time to sleep. To survive, one had to learn to sleep in the very few moments when it was possible.

And those memories led her to think about Sergeant Major Theodorus Wasp, now Lieutenant Wasp. They had grown close during the stress of combat; she had cried in his arms after the attack on the command center when she had lost half her crew in minutes. Mostly, she had been terrified at her own newly discovered lust for battle and at the thought she would never be the same. He understood. She felt no one else could ever really know what she was going through.

After they had returned to Earth, he volunteered for Officer Candidate School. She knew he had done it for her, because both of them being officers would make any relationship easier. It was a sacrifice, she knew. In the Marine Corps, Sergeants Major were said to be next to God. Second Lieutenants, though higher in rank, were not. If not for her, he would happily have served out his career as a senior NCO.

She had been at his commissioning ceremony, pinned on his new rank, even. He had looked the part. He was young for a sergeant major, in his early thirties. So, the second lieutenant rank did not look awkward on him.

After his commissioning, they had spent one long, glorious week together in a cabin in the wilderness of Canada. Lee had been sure in her heart that he would be her man for life. She had never been so happy.

And then he had gotten orders for some sort of classified assignment, and she had gone off to command school. And then *nothing*. No communication from him at all. She kept trying to find out where he had gone, but nothing turned up. Then she had a visit from two officers from the Marine Intelligence Agency. They made it clear she was not to inquire further about Lieutenant Wasp, and that doing so would potentially jeopardize his mission security.

And that was that. She heard nothing from the man she thought would be hers forever.

Alas, whatever Danner had cooked up for her could wait an hour, and she would be more receptive with a little sleep.

CHAPTER 10

CHAMBERS AND NEMETH (TWO YEARS PRIOR)

"Fly the *Centurion* into the path of the missile, then detach the shuttle before impact," said Rear Admiral Jay Chambers.

"Roger, sir," replied Master Warrant Officer Mathias Nemeth.

Chambers and Nemeth were in the shuttle, which was attached to and controlling the command ship, *Centurion*. Chambers had ordered the entire crew of the *Centurion* to abandon ship as he intended to ram the *Centurion* into the Cruiser *Okinawa* to prevent it from ramming the Moon Base Alpha 51, where his special operations team had just captured evidence of a quantum flux generator that had been used to attack Alliance targets across the galaxy.

Chambers' niece, Lieutenant Commander Mei Ling Lee, was commanding the special operations task force on the Moon. If the *Okinawa* struck the Moon, Lee and the team would be killed. Further, the evidence that Chambers had sought to acquire proving that the galactic war was a hoax perpetrated from within the Alliance itself would be lost.

In a strange turn of events, the *Okinawa* was commanded by Chambers' estranged son, Commander Jay Chambers, Jr. Father and son had not seen each other for many years. The younger Chambers was an unwitting accomplice to the plans of those behind the attack on the Alliance. Following extended combat between a force of two cruisers sent by the cabal within the Alliance and the Achilles Battle Fleet, the *Okinawa* had been severely damaged and was now stripped of its shields and armaments.

The *Okinawa's* Captain, the younger Chambers, had ordered all the crew of the *Centurion* to abandon ship and he was now flying it directly toward the moon base, intending to ram it. His own commanding admiral, an impostor clone, had told the young Chambers that the destruction of the moon base Alpha 51 was a *Primary Concern,* a term that meant the mission was to be accomplished at all costs.

Chambers and Nemeth sat side by side in the small shuttle which was anchored to the command ship. The navigation controls for the larger vessel were routed into the shuttle, so they were effectively steering it.

"Has everyone evacuated the command ship?" asked Chambers.

"Yes, sir," said Nemeth. "Ensign Danner reports one hundred percent are accounted for. He has the pods in a formation, and they are moving away from the combat area to the far side of the gas giant. They are as safe as they can be."

"He did well, don't you think?" said Chambers.

"He has potential," said Nemeth. "He did well under stress. His decision to be a person of substance lies in his future."

"You said the same about me when I finished the special warfare course," said Chambers.

"And I was right," said Nemeth.

"Time to impact?" asked Chambers.

"Seven minutes," said Nemeth.

"Hail the Captain of the Okinawa, please."

On the screen, appeared a young-looking naval commander. He had a mix of Asian and Caucasian features.

"This is commander Chambers of the Alliance cruiser *Okinawa*," said the young man.

"I am Admiral Chambers, Commanding officer of the Achilles Battle Fleet." said the elder Chambers.

The young man showed nothing, no emotion, *a trait he got from his mother*, thought the Admiral. Neither spoke for a long moment.

"You hailed me." said the younger man.

"Jay, I know you think you are doing the right thing," said the Admiral. "But you are being misled. Your Admiral Stuart was an impostor, a clone."

"And you have proof of this?" Asked the younger man.

"Of course," said the Admiral. "We have him in custody. We have his DNA. He is not the admiral. He's an enemy agent. We had clones here in the battle fleet as well. My own deputy. Many died because I didn't see it."

"And this moon?" asked the Commander.

"It houses a Macro Wave Collapse generator," said the Admiral. "It's why your clone admiral doesn't want it captured. Because it is proof of what they've been doing, that they're behind these attacks."

"I have my orders," said the younger man. "I was told the Moon's base was a *primary concern*."

"I figured that," said the Admiral. "Your orders are invalid and illegal."

The Commander said nothing.

"Don't sacrifice your crew," said the Admiral. "They deserve to live. Their deaths will mean nothing."

"They have all left the vessel," said the Commander. "I ordered them to abandon ship. It's just me."

"I have a team on the Moon," said the Admiral. "They have control of the generator, it's the proof we need. If you destroy it, they will have died for nothing."

"Admiral, you and I both know it's a special ops team. You have time to get them off," said the Commander. "It's your job to protect your people, not mine."

"Jay," interjected Nemeth. "This is Uncle Matthias, you know me."

"Yes, Uncle," said the younger Chambers with a voice that softened somewhat.

"Your father is speaking true," he said. "He does not lie. *I* do not lie."

"Duty is all, Uncle," said the Commander. "*You* taught me that."

"Your duty lies elsewhere, Jay," said Nemeth.

The young commander stared back, the beginnings of indecision on his face.

"There is more, Nephew," said Nemeth. "The Moon has a quantum flux generator. It is cooled by liquid hydrogen from the gas giant. If you hit, the moon will go, the generator will go, the giant will go, the star will go. None will survive."

No response.

"And on the Moon, there is one who is blood to you," said Nemeth. "The daughter of your mother's sister. Mei Ling Lee. Your cousin. She has fought with valor, suffering many injuries. She cannot, will not, leave her duty. Even now she is in great peril. Do not kill her, Jay."

"I will not surrender an Alliance warship," said the younger Chambers.

"Agreed," said the Admiral. "Divert from the Moon, pick up your people. Under truce, together we will lead a joint delegation to the Moon and see what we will see."

Just then, a missile was launched from the Moon's surface and headed straight for the *Okinawa*.

"Jay, that missile is for you. Your shields are down," said Chambers. "Eject!"

"Not today, Father," said the younger Chambers. "This is my command, and I won't leave it."

The connection to the *Okinawa* was broken and both Chambers and Nemeth watched on the operational display as the missile rose from the surface of the moon directly toward the unprotected *Okinawa*.

"Can we intercept the missile?" asked Chambers.

"Yes, sir," replied Nemeth. "Just barely, but..."

"Tell me," said Chambers.

"In order for us to detach our shuttle from the *Centurion*," replied Nemeth, "we will need to drop the *Centurion's* shields. The missile will likely cause catastrophic damage to the *Centurion,* and we will be within the hazard radius from debris, even if we detach now."

Chambers hesitated for just a moment. Nemeth looked at him with surprise. In all their battles over the years, Nemeth had never seen the old warrior hesitate in a crisis.

"Do it," said Chambers with resolve. "Get us clear if you can." Nemeth flipped the switch that dropped the shields protecting both the *Okinawa* and the shuttle, and performed an emergency disengagement, which was basically an explosive charge that sent the shuttle spinning away from the *Centurion* seconds before the missile from the surface plowed into the underside of the warship, just aft of the engineering section. The explosion was catastrophic. It basically broke the back of the *Centurion*, sending the two pieces falling toward the moon. Fragmentation from the blast flew out and struck the shuttle.

"We're hit," said Nemeth in his unearthly calm voice. "Stabilizers are out, ruptured fuel cell, shields inoperable."

"Can we make it to our rally point?" asked Chambers. From long experience with special operations, pilots always designated one or more rally points to go to in case the vessel was damaged, or the mission could not be accomplished as planned. Chambers and Nemeth had picked an abandoned underground hangar on the far side of the moon.

"Yes, sir," said Nemeth. "But the shuttle will be dead after that. It will need repairs before it can fly again."

"Well, it's not the first time that's ever happened," said Chambers with a smile. The two of them had served together as special operations warriors decades before. Chambers had been the squadron commander and Nemeth had been the squadron master chief petty officer. Together, they led dozens of combat missions. In one of those missions, Chambers had been wounded, lost his left hand and was brutally scarred over the left half of his body. After that incident, Chambers could no longer meet the physical demands of the special operations forces. He had the hand replaced with an artificial one and had kept a wicked scar on the left side of his face.

Chambers had been offered a full medical pension and had refused it. Instead, he changed over to be a logistician and rose up through the ranks to Rear Admiral. Only the attack of December 7th, in which all other flag officers had been killed, had thrust him back into a leadership role.

"Not the first, or second," said Nemeth with a rare smile.

"Shall I send a distress signal?" asked Nemeth.

Chambers paused and slowly shook his head, then looked directly at Nemeth. "We should go black, wait this out. We may not be able to take the fight to those behind this if we stay in the system."

"I agree," said Nemeth. "But Mei Ling will take the heat when you're not there. Are you up for that?"

Chambers breathed deeply, nodded as if to himself and then said, "Yes, I can live with that. Mei Ling is tough, she has resources, and let's not forget the Chinese government won't take kindly if she is made a scapegoat. Can we get an encrypted burst transmission to our friends at the Chinese consulate in Denver? That would give her some room to maneuver when she gets back."

"Yes, I can get a message through," said Nemeth. "We'll need to wait until this settles out. There will be a lot of attention here in the next weeks."

"Do we have any friends in the Marine Corps who might help Mei Ling?" asked Chambers.

"Well, our old friend, Motubu, is now the Commandant. He was a good man in a fight, he saved us on that operation on Rigel," said Nemeth.

"And we saved him, more than once," said Chambers. He turned to look at Nemeth questioningly. "You said he's the commandant of the Marine Corps? A four-star general? I wouldn't have thought he had the politics required for that. He's more a warrior, and a risk-taker."

"True," said Nemeth. "There was talk he wouldn't get it, that he didn't kiss enough ass. But he did get it. He is the first East African to hold that position. His combat record is unassailable. The rank-and-file love it. Do you want to reach out to him?"

"Yes," said Chambers. "I'll do it when the dust settles here."

Chambers turned to look directly at Nemeth. "And what about you? If we go dark, Laura won't know about you. She may think you died. You OK with that?"

Nemeth did not answer right away. After a few moments he said, "All these years I've been single. We've been on operations with married colleagues. Sometimes they have kids. We've been on black ops when no one outside the team knew where we were or when we'd get back. Families were told nothing. Do you remember Petty Officer Max Whalen? He had five kids. We couldn't even tell his wife he had been killed in combat for a year after he died because the mission was that sensitive. Even then, she couldn't be told the details. Whalen died a hero. He saved us. She wasn't told. So, no, I don't like it that Laura's kept in the dark, but that is the life we picked. I did promise her I'd return, and I will."

CHAPTER 11

REARM AND REFUEL

Nemeth used his flying skill to land the damaged shuttle in the abandoned bunker they had picked as a rally point. Using a caution borne of years of combat and survival, they explored first the hangar then the connecting tunnels. They mapped the routes, and then checked for structural damage, reinforcing areas that were damaged or simply worn out over the decades of non-use.

It took them weeks, but they finally found what they needed: supplies, fuel, weapons, and ammunition; most of it left when the mining enterprise had been abandoned about 80 years prior. Much of it was old but usable in some form. Chambers and Nemeth were special operations experts, trained to make use of anything when necessary. In addition, Chambers trained himself in every area of maintenance and repair when he took on his career as a logistician.

And it was a relatively good time for them. They worked well together, a trait they had developed years before. They didn't say much, and that was OK with both. They fell into a daily routine: collecting supplies and carrying them, sometimes for miles, through the tunnels;

cleaning everything off to see what was still usable, running function checks on parts; and slowly and carefully repairing the shuttle.

They had agreed that the shuttle would need upgraded capabilities for their next mission. Increased armor, upgraded shields, better offensive weapons and, most importantly, a faster-than-light or FTL drive. Finding parts for the FTL drive took the longest. They had to wait for the Alliance investigation team to finish their work and depart the Moon. Then they carefully worked their way into the technical areas where the work on the macro wave collapse had taken place. Much of it had been destroyed, but not all. Slowly, as weeks led to months, they got what they needed.

People who were not in special ops, often imagined special operators were cowboys who led exciting, devil-may-care lifestyles. Yes, there was a time of extreme exertion, where risk and highly honed skill came into play. But at a ratio of ten-to-one, most of a special operator's life was spent in preparation, training, endless rehearsal, and meticulous planning.

Every day, one would read off the checklist for preflight, while the other performed the check. When they found something, they stopped and fixed the problem, then started again.

The following day, they switched roles and repeated the procedure. They had learned from long experience, *never leave anything to chance*. It was more than just a safety issue. On this mission, there wasn't going to be any backup. They wouldn't be able to call for help if their FTL drive burned out somewhere in interstellar space. Everything had to work and, where possible, there had to be redundancies in onboard systems.

They also trained. Both men were past 50, they knew the years and the accumulation of injuries had taken their toll. They would never again be as fast or resilient as when they were in their 20's. Success and survival would depend on experience, training, timing, and judgment.

For their next mission, they would primarily need skill in close quarters combat. That meant small arms, knife fighting, and hand-to-hand combative skill. Every morning, they trained for general fitness for an hour. This included strength training, cardio fitness, and flexibility.

In the evenings, after a long day of gathering supplies or maintenance on the shuttle, they would train either on marksmanship or hand-to-hand combat. For the marksmanship, Nemeth had set up an automated pistol range, with pop up targets where the computer selected random timings and locations. The shooter had to hit the target or targets in rapid succession, and not engage a target that wasn't a threat. The computer also fired back, not with bullets, but with a line-of-sight laser designed to sting, but not injure.

If the shooter got hit, the computer in his body armor would 'decide' how bad the injury was. If it would have been fatal, the armor and weapon shut down and the shooter was practically 'dead' for that iteration. If it was only a non-fatal wound, the computer would incapacitate part of the armor to simulate the loss of that body function.

Nemeth, though older than Chambers, was the better shooter in this drill. Nemeth had been Chambers' special operations instructor when Chambers, as a very junior petty officer, had qualified for the special operations forces. Chambers had later gone to officer candidate school to become an officer. The two had served in the same units repeatedly over the years. When Chambers, at the rank of Lieutenant Commander, was selected to command a special operation squadron, Nemeth had served as the squadron's master chief petty officer.

During a hostage rescue operation, Chambers had been severely wounded, losing his left arm below the elbow, and had suffered serious burns to the rest of his body and face. Even after rehab and the addition of an artificial left hand, Chambers was unable to meet the demanding physical standards of the special operations. He was offered a medical retirement on full pension, but he turned it down.

Instead, he transferred to the logistics corps and set about learning a new profession. In that capacity, he had risen to the rank of Rear Admiral. Although Chambers had maintained his fitness, it had been many years since he had served in the demanding training environment of the special operation warrior.

Nemeth had no such impediment. After Chambers had left the special operations forces, Nemeth had become a qualified aviator and been promoted to warrant officer, but he had stayed in the special operations community and thus had kept and honed his skill over the years. He was deadly accurate and made all his moves with what seemed like blinding speed. Both Nemeth and Chambers knew it wasn't just speed. It was an economy of motion. Nemeth had long since mastered the art of short, quick, fluid movements, combined with a keen decisiveness.

After one of the drills in which Nemeth had both survived and *killed* all of the pop-up targets, the two of them watched the exercise on recorded video.

At one point, Chambers slowed down the feed, pointed to the screen, and said, "I don't get it. You began your turn toward that last target before it popped up. Do you know the algorithm the computer is using?"

"Yes and no," replied Nemeth. "I was turning toward the area where I would be most vulnerable. In this case, that's to my left rear. The computer knows I am less accurate and less aware in that direction partially because of a hearing loss on that side. But that's not why I turned that way. I did it because that's where *I* am more vulnerable."

Later, as they were going through the marksmanship exercise, Chambers had, once again, been *killed* during the exercise,

"You're dead," Nemeth said, in his dead pan Eastern European accent. "You are forgetting to move. You can't just stay in one place and hope to take out all the targets. The system's computer will always produce two or more adversaries at the same time if you stay still long enough. You have to move, pick up the targets, and fire while in motion. The bad guys don't care that you are an old decrepit Rear Admiral." This last statement was said with a smile, but the point was made.

In unarmed combat, Chambers was Nemeth's equal. Some 40 years ago, Chambers had trained under a grandmaster in Hong Kong. The man was one of the best, and perhaps the most demanding martial arts instructor Chambers had ever seen. Toward the end of the year, young Petty Officer Chambers had fallen in love with the grandmaster's young sister, and the two had eloped.

In that family, at that time, marrying a Westerner was considered taboo. Chambers had left Hong Kong with his young bride without the permission of her father. It caused a rift between her and the rest of the family.

After that year, Chambers kept training and found other masters to train with. That training, combined with the combative training he got from the special operations qualification, made him formidable indeed.

Chambers and Nemeth would square off with each other, wearing what amounted to full protective body armor. They would run through attacks and defenses, honing their skills, and adapting to the fact they were not as strong or as fast as in years prior.

The heart of Chambers' skill was the ability to move quickly, decisively, starting from absolute stillness. He gave no tell. It took years of training and practice to learn the art of complete relaxation, then a transition to explosive violence.

The strike needed to disable the opponent. The science behind it was the limitations on reaction time. The average person needed two-tenths of a second to react to anything visual. Chambers had trained himself to move and strike consistently in less than one tenth of a second.

As with marksmanship, it came down to decisiveness, timing, and judgment. At one point, Chambers struck Nemeth fully in the face guard before Nemeth could even raise his hands in defense. The force of the blow from Chamber's artificial arm knocked Nemeth backward, who rolled as he had been trained, and came up in a fighting stance.

Nemeth said, "I don't recall that technique in the special operations manual. Is that from Master Lee's bag of tricks?"

"Maybe," said Chambers with a smile. "I'll show you."

And that is how they worked for months until they felt both the vessel and they themselves were ready.

CHAPTER 12

AN OLD FRIEND

After six months, they took stock. They were in the shuttle having gone over a final checklist for readiness.

"The shuttle is ready," said Nemeth. "It has FLT drive, enhanced armor, shields, and a suitable suite of weapons including lasers, missiles, a cannon, and even a small rail gun. It has modified thrusters, so in non-FTL drive it can maneuver almost as well as a fighter. We have enough food, water, and medical supplies for months."

The two looked steadily at each other. Finally, Nemeth asked, "And where to now?"

Chambers pointed to a screen and punched some buttons. The screen showed a star system.

"It's TOI-700, the locals call it Toy-700 or just Toy," said Chambers.

"Why there?" Asked Nemeth.

"The regional government there has collapsed," said Chambers.

"Nothing special about that," replied Nemeth. "Much of the outer rim has fallen away from the Alliance since the attack of December 7th. Why this world?"

"Here's why," answered Chambers. He pressed some buttons and the screen changed to a video feed showing an incoming call. What appeared was a face they both knew.

"Admiral, Chief," said Chief Warrant Officer John Raymond. "It's good to see you both alive."

Nemeth looked at Chambers with raised eyebrows. Then looked back at the screen and said, "Dr. Raymond, it's good to see you as well. I take it you are no longer in the service?"

Raymond had served as the maintenance chief petty officer for the 514th Recon Squadron during the attack of December 7th on the Achilles Flotilla. For the 20 years prior to that battle, he had served as an enlisted person in the Navy. He was well respected and served with distinction. Unbeknownst to anyone, he was in fact Dr. John Raymond, one of the most famous physicists in the galaxy.

He had joined the Navy in response to what he believed to be covert government threats to his life after having worked on a classified project that held the potential for instantaneous transportation of large objects over great distances. For most of his career in the Navy, he had served aboard starships in a relatively low-profile career field in hopes of avoiding being assassinated by covert government operatives.

When the war started, and it became clear the mysterious adversaries were using a heretofore unknown technology, Raymond revealed his identity and worked with the admiral and the staff to understand the new technology and develop countermeasures. Those efforts had eventually led to the discovery of a macro-wave collapse generator on Moon Base Alpha 51.

Chambers had ordered a Marine Commando special operations raid on the Moon to seize the generator and preserve evidence of the use of the generator to sabotage the Alliance.

After the battle, the Achilles Battle Fleet, minus Chambers and Nemeth, returned to Earth. Evidence that the attack of December 7th had been an apparent attempted coup by Alliance insiders, masquerading as the ancient enemy, the *Others*, had thrown the Alliance into chaos. Attempts to arrest and bring to trial senior members of the Navy and the government charged with treason caused portions of the Alliance to attempt to secede. In the planetary systems at the limit of Alliance control, fighting had broken out as the secessionists sought to establish a foothold.

"You're correct, Chief," said Raymond. "I retired when I got back and have started a private research firm, *Raymond Research Associates,* or RRA."

"We saw that your return to civilian life caused a stir," said Chambers with a smile. "You've become quite a celebrity."

Raymond gave a rare wry smile and said, "That was unavoidable. I figured the best way to protect myself would be a full disclosure. Any information I had is out in the open. Anyone who wants to kill me now for revenge is going to have to do so in a very public way."

"I hope you are taking precautions," said Chambers.

"I am," replied Raymond. "My firm has its own security and I have taken reasonable precautions with my living arrangements and movements."

"I hear a *but* in there," said Chambers.

"You're right," said Raymond. "I know full well if a powerful entity wants to get at me, that can be done. All I've done is made it so that any such adversary would need to expend resources and take risks to get at me." He shrugged and said, "And now that the conspiracy is out in the open, I'm thinking they have no need to take me out. Their reason to remove me was that I knew about the macro-wave collapse process they were using to attack Alliance assets. Now that has been revealed, what would be their motivation to kill me? In the meantime, I just live my life."

"And how is your recovery?" asked Chambers. During one of the attacks on the Achilles Battle Fleet, a clone-hybrid disguised as a senior member of the staff, had attacked Raymond and seriously injured him. By the time the fleet returned to Earth, Raymond had only partially recovered.

"Good as new," said Raymond as he tapped his forehead with a finger. "When I got back, the Navy sent me to the medical center at Perth where Laura Zakany put together a world-class group of neurologists to check me out before my retirement."

At the mention of Laura's name, Nemeth broke his usually reserve and asked with animation, "How is she?"

"Healthy," said Raymond, "and when I saw her last, very pregnant."

CHAPTER 13

OUR LOVED ONES

Nemeth opened his mouth, then froze, his eyes wide. Chambers and Raymond seemed to understand that the fact that Laura Zakany was pregnant was news to Nemeth. Both had the sense of respect for Nemeth to stay quiet, letting him come to terms with the new information.

Nemeth breathed deeply and then said, "How was she, how did she seem?"

"She was as you would expect. She is tough, ferocious even," said Raymond. "She was enrolled in medical school. They put her in the third year because of her experience and qualification as a nurse practitioner." Raymond smiled and continued, "And she was pissed at the Navy for trying to declare you dead. They sent a priest and an officer to make the notification. She threw them out. Called them idiots. She is convinced that you are alive. I'm glad to see she was right."

The three were quiet for a long moment, then Chambers said, "And Mei Ling?"

"When she got back to the Denver spaceport, the Navy tried to have her arrested," said Raymond. "The Chinese ambassador intervened and gave her sanctuary at the Chinese consulate in Boulder. The rogue elements in the Navy sent a clone force to kidnap her, but Wasp, with a squad of Marine Commandos, sorted that out nicely. General Motubu arranged for her to transfer to the Marines and then convened an impromptu court martial at which she was acquitted of all alleged crimes."

"And now?" asked Chambers.

"She is reportedly doing well in a series of courses designed to prepare her for company command."

Chambers took a breath and visibly relaxed. "Any word on my son, Jay, Jr.?"

"That's a little bit cloudy," said Raymond. "After the battle of Alpha 51, there was a huge investigation, public outcry, and such. Within the rank and file in the Navy, Commander Chambers got major respect for his leadership during the battle. He didn't surrender, he protected his crew by having them abandoned ship, and he stayed with the *Okinawa* even when it became clear it would be destroyed."

Raymond paused and then continued, "On the flip slide, he *was* on the wrong side of that battle, and he took orders from a fraudulent admiral. That said, his involvement got overshadowed by the disruption to the Alliance and the Navy when the truth about the attack got out. The Navy split, some going with the separatists, some staying with the Alliance. I don't know which side he went with. As far as I know, he is still serving. I can say for sure, there was no public discipline or court martial."

Chambers showed no reaction to this news, just nodded and said, "Thanks for the update, Chief. Now tell us what you found out about TOI-700."

CHAPTER 14

MISSION PLANNING

Nemeth looked at Chambers with a quizzical look.

Chambers nodded and said, "Some months ago, I reached out to the Chief and asked him to do some research for us."

Nemeth shook his head and said, "Good to know."

Raymond nodded, tapped some keys, and the screen transitioned to a three-dimensional portrayal of a star system. It showed a star with five planets.

He said, "This is the TOI-700 system. The star is a red dwarf, spectral class M. It was first discovered in the early twenty-first century. Astrophysicists of that time paid attention to it because they believed at least one of its planets, then called TOI-700 D, later renamed *Ulysses*, orbited in the habitable zone. Once FTL was developed, it was one of the first systems investigated for possible colonization."

"What's so special about it?" asked Nemeth, somewhat sharply.

"Give me a moment to give you some background; it will make sense in a bit," replied Raymond with irritation. Then he looked at Chambers and said, "You did ask me, and I've spent some considerable resources and time getting this information."

"Please continue, Chief, at your own pace," said Chambers.

"Initial enthusiasm for the Ulysses colony faded once it became clear there were thousands of planets suited for colonization within range of the FTL drive. The colony that was set up there sort of sputtered. Within a hundred years, it was down to a few hundred thousand people and was considered a backwater. It's not on any of the major trade routes. The Alliance Navy makes a port call with logistic vessels every few years, bringing necessary supplies, mostly medical and some technical items that can't be manufactured on the planet."

"So, what's changed?" asked Chambers.

"When the Alliance started to break apart, there was a substantial increase in both vessel and communications traffic."

"How is that an outlier?" asked Nemeth. "Many of the planetary systems got taken over by the separatists. Wouldn't that explain the increase in activity?"

Raymond nodded and thought for a moment before replying. "Yes and no," he said. "Yes, in that if the separatists had gone there, there would have been an increase in traffic of all kinds. But no, that doesn't seem to be what has happened." Raymond stopped and looked as if he was thinking of something.

"And...," said Chambers.

Raymond looked like he had been startled and said quickly, "No rebel vessels are known to have gone there. And the separatists, in every other instance, made a play for valuable star systems, those on trade routes, or with substantial industrial or agricultural capacity. None of the backwater systems like Toy were taken."

Chambers looked steadily at Raymond and said, "Chief, something there has your attention, what is it?"

"OK, so let me tell you what we do know, and you'll see the anomaly," said Raymond. "Like many of the outlying systems, it is believed the government collapsed. By itself that isn't unusual. Many of these smaller worlds have fairly relaxed systems of governments, often based on tribal affiliation, custom and family interactions. We wouldn't really consider it a government at all. There's a tribal chief who settles disputes and appoints judges and administrators. They have general elections every four years, but these are often pro forma affairs, more of a social event. There's a festival, a feast day, parades, and speeches. Citizens do vote, but they don't take it seriously. Voting day is mostly a festival, a recognition of their long history of a free democracy. Big changes are rare."

"So, the government was institutionally weak and it collapsed," said Chambers. "Where's the issue?"

"It's just that," said Raymond. "The relaxed type of government, one based on long-held traditions, should have remained stable. It shouldn't have collapsed, unless there was some outside influence that caused it to cease working. The Alliance was already pretty much ignoring the place. Its fragmentation should have had zero impact on the planet."

"Granted," said Chambers. "There's a minor mystery about the government ceasing to function. What did happen?"

"Like so many other systems, Toy fell into disorder and rival crime gangs rose up vying for control, mostly of the single large city, Ephesus."

"Any anomaly in that?" asked Chambers.

"Yes," said Raymond. "I have a team of sociologists on my staff. They say something is off. Crime gangs typically fight over the control of the production and distribution of contraband. That's because contractual relations for contraband can't be enforced by the government or the courts. Because of that, these relations are enforced by the threat of extreme violence. Nobody fights over property that's already legal."

"So, what's the contraband on Ulysses?" asked Chambers.

"That's just it; there isn't any contraband that we know of," replied Raymond. "Ulysses has a barely self-sustaining economy. There's no evident contraband and if there was, there isn't enough excess income to make it worth anyone's effort."

"What else?" asked Chambers.

"There's some indication that the rival gangs are from off world, not native to Ulysses or the other inhabited planets in the system," said Raymond.

Chambers got a far-away look that indicated he was thinking about something. Both Raymond and Nemeth remained silent. They had served under the admiral for long enough to know he wanted time to think over what he had heard.

Finally, Chambers said, "OK, I get it. The logical assumption is that something in the system is, or has become, valuable enough to draw in a criminal element from outside the system. I assume, Dr. Raymond, that you are going to tell me what it is."

"I will tell you, but first I need to give the background that will make what I'm about to explain understandable," said Raymond.

Nemeth gave an audible snort showing frustration for Raymond's lecturing style. Raymond was infamous for his patronizing, sometime condescending methods. Nemeth, as a special operations warrior, was always after actionable intelligence, not intellectual rambling.

Both Chambers and Nemeth recalled how Raymond's lecturing style had irritated Lee during command briefings when she was the Battle Fleet's chief of staff.

But Chambers stayed patient and said, "Please continue, Doctor."

Raymond looked hurt by Nemeth's impatience and waited a long moment before continuing.

"You will recall that during the attack on the Achilles Battle Fleet that took place during our attempt to snatch one of the macro-wave collapse vessels, our pilots encountered a slightly larger vessel, that they termed a *corvette*."

Nemeth said, "Yes, we fought it. When it blew, it had an enormous yield, far more than a nuclear warhead or a fusion drive."

"Right," said Raymond. "We later learned that the reason the enemy had to use such a vessel was because the power required could not be projected across interstellar space from the Moon at Alpha 51 where the MWC generator was located."

"Yes, we knew that", said Chambers. "The generator was too far away, and they needed a platform to project power."

"The technical problem wasn't distance, per se," said Nemeth, becoming animated as he got closer to his main point. "It was the sheer complication of relative location and timing of spatial events so far removed. The system that contained Alpha 51 was moving away from our fleet at relativist speeds. Trying to influence events at that distance is too complicated because the two systems are not in the same frame of reference."

"And because the energy signal was so high," said Nemeth. "The MWC generator needed to be well away from traveled areas. Otherwise, it would have been too easily detected."

Chambers frowned and then said, "It was a limitation on the enemy's ability to project combat power. They either needed an MWC generator within the system where they were trying to direct forces — something that would be easily detected and taken out — or they needed a suitably equipped vessel, in this case a corvette, to handle and direct that much power. In either case, the enemy had a built-in vulnerability. We take out the generator or the command vessel and the enemy cedes all advantage."

"Right," said Raymond. "That's why they pretended to be the *Others*. Why it was done as a surprise attack? It was designed to get intelligence services to look to *Others'* space for any threat, not internal space controlled by the Alliance. It was always a gamble."

The *Others* were a mysterious non-human race that had waged war with the Alliance over one hundred years before. When the attack began, most people assumed it was the *Others* recommencing their war with humans.

"For God's sake, John" said Nemeth, finally losing his patience. "Please tell us what you've found!"

Chambers remained calm, but he didn't correct Nemeth. The time had come to tell the story.

CHAPTER 15

ASTEROID MINING

"The issue for the separatists is control, specifically control of the MWC effects at a distance. As I said before, the distances are just too great under existing technology," said Raymond.

"How does what they're doing in the Toy system solve that problem?" asked Chambers.

"Short answer, *quantum computing*. Long answer, way more complicated," said Raymond.

"But we already have quantum computers," said Nemeth. "I thought our FTL drive depended on their computations to pick safe routes for our jumps?"

"That's true," said Raymond. "But our current quantum computing is relatively limited. The technology requires special materials that are not generally available, and the demand for quantum computing is so great *because* so many military and private entities need it."

"What's on Ulysses that they need?" asked Chambers.

"It's not on Ulysses; it's in the system's Oort Cloud," said Raymond. He stopped speaking and let that sink into his audience.

"They're mining it," said Chambers simply.

"Yes," replied Raymond. "Quantum computing requires special materials. Specifically, what are termed *Van Waals* materials."

"Van Waals?" asked Nemeth.

"The problem with quantum computing is that it requires material that is extremely thin, only an atom thick. Only a few materials are known to be able to achieve this property, the so-called 2D or two-dimensional materials. Mainly, this is Boron Nitride or more properly Hexagonal Boron Nitride or hBN."

"Is there a shortage of this hBN?" asked Chambers.

"Boron Nitride is artificially manufactured. It is rarely — very rarely — found in nature. It is expensive and its production is restricted by cost, so it is relatively easy to monitor," replied Raymond.

"Meaning, if the separatists wanted to hide its manufacture or purchase, they wouldn't be able to do it," said Chambers.

"Correct," said Raymond. "They couldn't do it easily or completely. If we looked hard enough, we'd see signs of it."

"Tell me about the mining," said Chambers.

"The mining activity explains the increase in vessel traffic in the system," said Raymond. "Whoever is doing it is using the space ports on Ephesus, but it seems mainly for refueling and maintenance, perhaps to rotate crews for the mining work, the main effort is out on the perimeter of the system."

"In the Oort Cloud," said Chambers.

"Yes," said Raymond. "As is the case with Earth's sun, Toy has a cloud of icy planetesimals that surround the star at approximately a distance of 3.2 light years, which is about half a light year deep. It's mostly icy comets, but there are enough rocky objects — asteroids — to make mining possible."

"And there are 2D materials in the Oort Cloud?" asked Chambers. "Why hasn't this been done before? 2D materials are in demand already?"

"Two possibilities," replied Raymond. "Either it wasn't economically feasible, or the ore found in the Toy Oort Cloud is pure enough to have induced this effort. The third possibility is that it is simply a way to get the materials they need, in the quantity they need, without entering the galactic market for 2D materials."

"It might be all three of those," said Chambers. "Chief, you are frowning. What's the problem?"

"There's more," said Raymond. "A newly found abundance of rare 2D materials doesn't solve the problem they were facing. Not even the most powerful quantum computer we now have, one that runs a billion times faster than our fastest super computers, could solve the spatial problems of distance manipulations across galactic scale."

"So, what are you saying?" asked an increasingly irritated Nemeth.

"Solving the problem of control at great distances isn't just about solving complex equations; it's limited by the lack of information."

"Explain," said Chambers.

Raymond thought for a moment, and then said, "Here's an example: take Earth and TOI-700. They are 100 light years apart from each other. If the MWC generator is on Earth, for example, the light arriving from Toy is 100 years old. So, if there was a supernova in the Toy system, a controller on earth wouldn't even know about it for 100 years."

"I see the problem," said Chambers. "A controller in a distant system can't make realtime decisions on how to employ MWC devices."

"So why the big, covert effort to mine these rare materials if having them doesn't solve the problem?" asked Nemeth.

"That's just it," said Raymond with what looked like real frustration. "We don't know. I've got some of the best theoretical physicists anywhere working on my team. We just don't know."

The three were quiet for a several moments. Then Chambers said, "The answer is in that system. That's where we need to go."

CHAPTER 16

MAKING ONE'S PRESENCE FELT

The bar was dark, smoke-filled, with low music playing. It was past midnight, but in this part of the city patronage was still strong and would be for another few hours. The tables and booths were about two-thirds filled.

Chambers and Nemeth had been on Ulysses for a month. They had built a cabin in the adjacent hills and had been making improvements and conducting reconnaissance of the city of Ephesus, the space port, and the local industry.

They had selected the bar, *Traveler's Home*, for a good reason. It was the de facto headquarters and hangout for the local crime boss, Anton Antonelli. His gang, such as it was, ran illegal intoxicants, mostly star dust, a central nervous stimulant, as well as under-age prostitution. It appeared to be a small-time, petty operation, really. But as they looked closer, Chambers and Nemeth could see it was a front for something else. There was more money, more people, and more activity than could be accounted for by the drugs or girls.

A week earlier, they had been in their cabin, discussing the next steps.

"I want to start with the bar," said Chambers.

"Because?" replied Nemeth.

"Because it's as good a place as any to kick the hornets' nest," said Chambers. "If it really is what it purports to be, we won't get much of a response. If it's what we think it is, a hub for the illegal mining, we'll get a different response."

"And do you want a full response from them?" asked Nemeth.

"Of course," said Chambers with a smile.

Nemeth smiled as well, and said, "We'll be ready."

"There is something else," said Nemeth. "I know you."

"I don't like that they are running children there," said Chambers with his war face. "Some of those girls are in their early teens."

"Before these outsiders came," said Nemeth, "this was a very traditional tribal society. These girls are local, taken by force from their families. The local men won't patronize the bars because of it. The customers are all foreigners, brought in for the trade."

Chambers nodded and said, "We can't solve every problem, but that's a start."

Chambers entered the bar. He looked like a local commoner wearing work clothes, and there were no others of that kind in the bar. Patrons stared. His scarred face and his artificial left hand stood out. Also, there was something off about him. He was old, maybe 60, but his muscled frame and the easy, confident way that he moved and looked about, seemed the deportment of a much younger man.

Chambers sat at a small unoccupied table near the side of the bar, his back to the wall. A too-young waitress, provocatively dressed, came up to him and said nervously in a whisper, "Sir, really, not the best place for you. Some bad men here."

Chambers looked at her without expression for a long moment in a way that made her fidget. She kept looking out of the corner of her eye, inclining her head as if to draw his attention to something in the bar.

Finally, Chambers said easily, "Could I have a glass of soda water, and could you ask the proprietor to stop by and see me?"

The young lady's eyes went wide, her mouth gaped open and then abruptly shut tight. She turned and fled without a word toward the back of the bar.

Chambers breathed deep, concentrated, listened. It was something he had learned from Grandmaster Lee, Mei Ling's father and his brother-in-law, many years ago. It was a sort of meditative trance. He heard, no felt, his surroundings. The sounds of breathing, the smell of the other patrons, the clink of glasses, the shift in conversation since the waitress fled. He was refining the information he had already gained when he entered the bar. He knew who was armed, where each person was relative to him.

A man stood before him, 30-something, tall with a bit of a gut but muscled in the shoulders and upper arms. Chambers looked up at him calmly.

The man said, "Gramps, time for you to go."

"I asked to speak to the owner," said Chambers calmly. "You are not him."

"All right, then," the man said as he stepped toward Chambers. "You..."

He didn't finish the sentence. Chambers sprang up, lightning fast, put the much larger man in a one-handed choke hold using his artificial left hand. The shock of the move pushed the man back a few feet. He was on his tiptoes, struggling to breathe.

In a flash, Chambers had drawn the struggling man's sidearm from the man's waistband. Still holding him, Chambers pivoted and shot a man at one of the nearby tables who had drawn his own weapon and was just rising to his feet. The bullet went straight through the man's temple, his head shot backward, and blood and brain matter sprayed rearward. Every person in the room stopped what they were doing. The armed men in the room froze, none wanting to be the next victim of the old man.

Chambers turned his attention back to the struggling man he was holding in his vice-like grip.

"You misunderstand me, young man. I asked to see the proprietor. Please get him." With a snap of his wrist, Chambers pushed the man away from him with great force. The man flew backward, arms pinwheeling wildly. He slammed into the bar about ten feet way, then slumped to the floor gasping for breath.

Chambers took a step toward him, and the man scrambled to his feet and said, "I'm going!" He held up a defensive hand and scampered toward a door leading to a back office.

Chambers sat back down, placed the handgun on the table in front of him, and signaled to the terrified waitress to bring his order. She came over, placed the drink on the table with trembling hands.

Chambers said gently, "What's your name?"

With quivering lips, she said, "Cherry, sir."

"That's not your name," he said simply.

"They said I have to say my name is Cherry." She paused as if considering how much to say. "My name is Amina." Then she blushed furiously.

"Amina," said Chambers. "It's a lovely name. It means truthful, yes?"

"And it means *devoted*, kind sir," she responded.

Chambers saw three men approaching the table from behind her. He said to the girl, "Amina, thank you for the drink. Best that you go now."

Amina nodded and hurried away, her head down.

Three men stood before Chambers. The one in the center and slightly forward of the other two was slender, below average height, black hair slicked back, a thin mustache. That was the boss, Chambers knew, Anton Antonelli.

Chambers assessed the two on either side of Antonelli. These were the muscle. From the posture and demeanor, Chamber figured they were a step up from the man he had so easily handled a moment earlier. Clearly, the two were armed, but their weapons were holstered beneath light jackets, hands at their sides, relaxed. Definitely a higher class of enforcer.

The man spoke in a soft voice, with a slight accent Chambers couldn't place. "Sir, you've caused a mess. I'd like you to leave."

Chambers looked directly at the man and said nothing. He let the moment age, showing no sign of nervousness or impatience.

Antonelli began to fidget. The old man wasn't frightened and wasn't an amateur. Antonelli seemed to realize that if this confrontation came to violence, he would die.

Finally, he said, his voice cracking against his will, "You asked to see me."

Chambers waited another long moment and said, "Mr. Antonelli, I did ask to see you. I am sorry for the mess, but your man was rude. And the other," he glanced toward the dead man on the floor, "reached for his weapon. I came here unarmed, I politely asked to see you, and I was treated with disrespect."

At this the man's eyes widened, and his face went pale. The two muscle men stirred, looking flustered for the first time. Chambers' mention of 'disrespect' indicated he was connected to the organized crime culture. If this old man was a high-ranking person in *any* organization, and Antonelli's man had laid hands on him, Antonelli was potentially in serious trouble.

"I apologize for the mistake, no disrespect was intended," said Antonelli. "How can I be of service?"

"I have come to make a change here," said Chambers. "You will now work for me. I will finish my drink and leave. Tomorrow, when I return, these girls will be gone, returned to their families along with the appropriate *Diya.*" A Diya was the traditional financial compensation in cases of murder, bodily harm, or property damage.

"You will at that time report to me the financial status of this establishment," continued Chambers.

Chambers stopped, said nothing more. He knew the best way to negotiate with power was to speak only what he meant, and to let that be all.

"Sir, I don't own the Traveler's Home," stammered Antonelli. "I can't just turn it over."

Chambers said nothing, just looked in the man's eyes. Then he made a gesture indicating that Antonelli and his men were dismissed, and he looked off into the distance and drank his club soda.

Later that night after Chambers had returned to the cabin, Nemeth asked, "How did it go? Do you think they will come?"

"Yes, they have to. They can't let that stand," said Chambers. "By now they will have assured themselves that even if I am connected to organized crime, I was there in the Traveler's Home without permission. They'll assume they have authority to take me out."

"Were you followed back here?" asked Nemeth.

"Of course," replied Chambers. "They aren't professionals, no subtlety."

"Think they'll come for you tonight?" asked Nemeth?

"Yes, no later than dawn," said Chambers.

"We are ready," said Nemeth with an evil grin.

And they did come.

CHAPTER 17

PREPARATION

Chambers and Nemeth had set up their cabin months ago. They had picked a spot in the hills outside Ephesus. Traveling the two miles to and from the city, they always took circuitous routes so as to throw off anyone trying to track them. That is until Chambers returned from the confrontation at the bar. Then he had come directly to the cabin, knowing he would be followed. Chambers and Nemeth had spent a good deal of time preparing for their visitors, whom they now expected in the hours before dawn.

The men who were coming were probably good in a bar fight, maybe. But they were not skilled infantry fighters. They would rely upon their advantage in numbers, not realizing that such an approach would actually make them more vulnerable. Nothing so awkward as a dozen men, clambering up unfamiliar terrain in the dark.

The route the men coming for Chambers would take had been well prepared by Chambers and Nemeth.

"We are ready," said Nemeth a second time.

"I don't want to just stop them," replied Chambers. "We need to send a message."

"Understand. Do you want me on the tail end?" asked Nemeth.

"That's best," said Chambers. "You're better in the dark, and with a knife."

"And the gifts we left in the city?" asked Nemeth.

"All set," said Chambers. "Every warehouse and port is wired. Because of the late hour, collateral damage should be minimized."

"I'll get going then," said Nemeth in his typically stoic manner, as he left the cabin.

Chambers turned to his computer console and flipped through the various camera settings. His screen displayed over a dozen different images in various filters of the route the men would take. They were ready.

CHAPTER 18

AN INAUSPICIOUS START

Anthony Moscatelli was excited. He had just turned eighteen, and it was the first time his uncle, Francisco, had allowed him to come along on an operation with his crew. Francisco was one of three sub-bosses reporting to Mr. Antonelli.

Anthony had no idea what this was about. He had heard there had been an incident at the bar, that one of Mr. Antonelli's men had been shot and killed. Shortly after that, Francisco had received an urgent call, and then had sent out an emergency recall for his team.

Anthony had never seen such a thing happen before. Normally, operations were planned days in advance. And he had never seen ten men on any single operation. The normal target for an operation was a shop owner, or farmer and his family. This never required more than a few men.

He could feel the excitement. The men on his uncle's team were experienced, tough men. Killers to be sure. They had been called to this mission unexpectedly, and he could tell by their demeanor that some had been sleeping and others smelled of alcohol. In the back of the truck, they were mostly quiet, the few conversations were hushed, with only a rare curse when the truck hit a bump.

The truck stopped and someone came around to the back of the truck and lowered the gate which gave off a loud squeak and a bang as it fell open.

"Damn it, let's keep it quiet," he heard someone say.

The men piled out of the truck onto what Anthony could see was a narrow dirt road with trees on both sides. It was dark enough that Anthony could only see the outlines of the other men. He reached for his holstered sidearm on his right hip, relieved to feel its weight. He had only a very quick orientation on the weapon and had fired it just once at one of the target ranges his uncle used.

"All right, gather around," he heard his uncle say. The men made a semi-circle around his uncle on the road. Anthony stood his ground when someone tried to shoulder him toward the back. This was his first op, and he wasn't about to be sidelined.

"The target is a man, about 60 years old," said his uncle. "He is known by a visible scar on the left side of his face and metallic artificial left arm."

"What the hell," said one of the older men. "Ten of us to take one man. Let me and Ernesto do it. Send this gaggle back to bed."

Francisco looked hard at the man who had spoken. He remained silent and men began to shuffle nervously. It was really bad form to question the commander on an op, Anthony knew. The speaker was Giuseppe Coscia. He was a senior *made* man. Technically, he was subordinate to Uncle Francisco, but it was known that he was a favorite of Mr. Antonelli from way back. Francisco would have to tread carefully with Giuseppe.

Finally, his uncle spoke. "Giuseppe, my brother. Thank you for the offer to handle this yourself. I have no doubt it would be done well and quickly. But I have orders from Mr. Antonelli. The target is worrisome. He was disorderly at the bar tonight. He killed one of Mr. Antonelli's men and made some inappropriate comments directly to Mr. Antonelli."

"All the more reason to let me handle it," said Giuseppe. He drew a twelve-inch knife from his belt; it gleamed in the moonlight. "I'll bring the boss his head."

Francisco nodded and said, "That will be your honor. You can take point with Ernesto. Where is Danny?"

"Here, sir," said a voice from somewhere in the group.

"Danny tracked him to his cabin earlier tonight," said Francisco. "He will lead us to the cabin. Giuseppe and Ernesto will handle the man. The rest of us will be in support."

"Who is this guy?" came a voice from the group.

"We don't have a name," said Francisco. "He has some skills, that was clear, or he'd already be dead."

"Is he a friend of ours?" asked a voice. The term *friend of ours* was understood to mean the man was a member of a crime family.

Anthony could feel the tension in the group at that statement. If the target was a member of any of the crime families, even laying hands on him could come with repercussions.

Conflict between the crime families was rare, but when it did occur, it entailed extreme violence. Every man present knew if this guy turned out to be a made man from any of the families, neither Mr. Antonelli nor Francisco would be able to protect those who took part.

"We don't think so," said Francisco. Anthony heard groans from the group at this expressed uncertainty. Francisco held up his hands in a placating manner and said, "Mr. Antonelli inquired and none of the families acknowledged him as a member. If he is a friend of ours, he's here without permission. The target is approved. Mr. Antonelli stands by it, and I stand by it."

Unnoticed in the darkness, Nemeth had joined the group. He was gratified that the group was beginning to doubt the wisdom of their venture. Before this was over, they would come to doubt it even more.

"Let's get going," said Francisco. "I want to be home by breakfast. It's about a mile to the cabin. Danny, you're in the lead with Giuseppe and

Ernesto. The rest in the rear as backup. And keep it down. We want to take him while he's sleeping."

CHAPTER 19

FROM BAD TO INSANE

It was an awkward start, thought Anthony. First, Danny, who supposedly followed the man back to his cabin, couldn't find the path that led from the road up to the narrow valley toward the target. Everyone got in a ragged line behind Danny, and when Danny couldn't find the path, they all followed him, while he searched up and down the road. Anthony thought they looked ridiculous, like a disoriented worm in the dark. Surely, this wasn't how operations were conducted. He was also embarrassed for his uncle, who was supposed to be in charge of this raid.

Finally, after about ten minutes, Danny found the opening in the trees, and they all followed him up the narrow path. One result of the delay was that the group had ceased trying to remain quiet. There were many grunts of frustration, and even a few conversations in normal speaking voices, way louder than Anthony thought was appropriate for what was supposed to be a covert operation.

Anthony was in the approximate middle of the file. It was dark, and the path was rough with roots and rocks. He tripped several times, and once fell hard and scraped his knees and elbows painfully. He could hear the ragged breathing and cursing of the men in the column. Most of the men were middle aged and apparently rarely did any exercise.

Then things went insane. Anthony heard a scream, something worse than anything he had ever heard or even imagined. It was the voice of Giuseppe, and he was clearly terrified. "Oh, God, no. *Please* no, no, no... Agheee!"

Everyone on the path froze, then something hit Anthony hard in his chest, and dropped to his feet. The object was heavy and wet, about the size of a basketball. It took him completely by surprise and it knocked him back a step. He looked down in the dim moonlight and saw what it was — a human head. It was Giuseppe's face, bloodied lips pulled back in a gruesome smile, one eye gouged out.

Anthony opened his mouth to scream, but just then there was a series of bright flashes coming from every direction accompanied by a loud screeching sound. The effect was complete disorientation. He felt a disabling nausea and dizziness. He fell to his knees and vomited.

Through the fog of his agony, he heard shouts and gunshots. Total confusion reigned.

From his rear, he heard a gagging sound and he turned to see the man behind him holding both hands to his throat, blood pumping through his fingers. Then the man fell on his face and began twitching.

Anthony reached for his weapon, but before he could draw it, something hit him hard in his face and he felt and heard his nose breaking. As he fell back, he felt someone move past him toward the next person in line in front of him. Then, again, he heard that sickening gagging, then silence.

Dear God, thought Anthony. He had never been so frightened in his life; never imagined one could be so frightened. He felt a warm wetness in his crotch area and realized he had just pissed himself. He decided to live if he could. He lay face down and tried to remain completely still. He wanted whoever this monster was to believe he was dead.

Anthony continued to hear shouts and gunfire, but these sounds faded and, in a few moments, all was silent. *This is it,* thought Anthony. *If I can just stay still, I'll live through this. I don't care what anyone says about it, I'll be alive.*

Then he heard it. Footsteps. Anthony prayed hard, *let this pass me by, please God.* The footsteps paused in front of him. Anthony willed himself to stop breathing. Maybe the man was just checking for those still alive and would think he was dead.

"It's OK, Anthony," said a voice in the dark. "You are alive for a reason. Get up."

The voice had the air of command, and Anthony complied. He pushed himself to a kneeling, then a standing position. He looked up and saw an older man, grim faced, long scar on the left side. The man was splattered in blood.

Anthony stood mute. If this man wanted to kill him, he would do so. Just then he remembered he was armed, but he was far too frightened to reach for the weapon on his hip.

As if reading his mind, the man said, "Don't bother reaching for the weapon. If I thought you could harm me, I would have taken it from you." There was a pause and the man continued. "Anthony, I've left you alive for a reason, and I've left you with your weapon as a sign of respect. This operation was a fools' errand, and I don't think you are responsible." The man's voice was mesmerizing, quiet, yet powerful.

"I need you to take a message back to Mr. Antonelli," said the man with the scar. He picked up the severed head by its hair and dropped it into a leather bag, then drew the opening closed with an attached drawstring. He handed the bag to Anthony, who took it.

"Take this to Mr. Antonelli. Tell him that his disrespect had consequences. Tell him, I will leave the bodies by the road so that their families can give them a proper burial." The man paused and said, "Do you understand?"

Anthony squeezed out a timid "Yes."

The man frowned and said, "Are you sure that's your answer?"

Anthony began to panic. What had he said wrong? He didn't want the man to kill him. He didn't want to die on this path in the woods.

Then Anthony said, "Yes, *sir*," emphasizing the honorific. "I will take the bag to Mr. Antonelli. I will give it to him and tell him you said that Mr. Antonelli's disrespect had consequences, and that you would leave the bodies of the men by the road so that they can have a proper burial by their families."

"You need to do that tonight, before dawn. I'll be watching. Please don't disappoint me." Then, in what seemed like an impossible movement, the man disappeared into the darkness.

CHAPTER 20

AFTERMATH

Back at the cabin, Chambers and Nemeth washed off the blood from their hands and faces.

"That was grim," said Chambers.

"Yes," said Nemeth as he shrugged. "It has been a while, at least for you. You've been pushing supplies these past two decades. Some of us have maintained the tradition."

Chambers nodded. His friend's comment was true. He *had* been working in logistics for many years. And the truth was, he did not enjoy the killing, then or now.

"You left the boy alive," said Nemeth. It was a statement, not a question.

"Yes," replied Chambers. "He's an innocent. And I do need someone to deliver the message. Let's see how he's doing with that."

Chambers turned to a computer monitor, pressed a key and they saw an image from the micro-camera that Chambers had placed on the front of Anthony's shirt during the ambush.

The screen showed Mr. Antonelli, hair disheveled, eyes burning with anger or horror. "What the hell. They're all dead?" he shouted. "Giuseppe too?"

"Yes, sir," replied Anthony. He noted, but did not point out, that Mr. Antonelli had mentioned Giuseppe, but not Anthony's uncle, Francisco, who was supposed to be in charge of the mission.

"How many were in the group that attacked our men?" asked Antonelli.

"I only saw one," said Anthony. "The man with the scar."

"That's bullshit," shouted Antonelli. "No single man could take out an entire team."

Anthony remained quiet. He only wanted to deliver this message and get out of there before Mr. Antonelli lost his temper.

"What was the message?" asked Antonelli.

"He said, 'Your disrespect had consequences.' At this, Mr. Antonelli's eyes went wide with alarm. "He also said that he would leave the bodies on the side of the road so that their families could give them a proper burial."

"How did you survive?" asked Antonelli with a sneer. "Did you run away?"

"No," answered Anthony honestly. "I couldn't run. I would have. I did try to play dead, but he caught me out."

"What's in the bag?" asked Antonelli.

Anthony handed the bag over and said, "He told me to give this to you."

Antonelli looked in the bag. His face paled, he looked up at Anthony and screamed, "I will kill this bastard with my hands. I will..."

Back in the cabin, Chambers pressed another button, and they heard through the video monitor a series of explosions.

Antonelli looked around wildly.

"And there goes the warehouses and port facilities," said Chambers. "Let's see how well I am received tomorrow night when I return to Mr. Antonelli for my report."

CHAPTER 21

PRESENT DAY: AN UNEXPECTED CHALLENGE

Mei Ling Lee sat strapped into her seat onboard the Navy shuttle. As she woke from her sleep, she heard the communication traffic coming from the cockpit where two Navy petty officers were flying the craft. She kept her eyes closed as she analyzed what she was hearing.

Something was wrong.

Lee was now a Marine Officer, but she had spent the first years in service as a Naval officer, rising to the rank of lieutenant commander.

"Shuttle Echo Four Five, this is Destroyer November Seven Seven, heave to and prepare to be boarded."

"Destroyer November Seven-Seven, we are an Alliance warship. Per intergalactic code section four five, we are not subject to boarding by non-Alliance vessels."

Lee snapped fully awake. For a non-Alliance vessel to attempt to board an Alliance warship was an outrageous breach of Naval protocol. And who had destroyers beside the Alliance?

She looked across the aisle and saw that Danner and Zakany were still asleep strapped into their chairs.

Lee unbuckled from her seat, walked up to the cockpit, and took a knee between the pilot and copilot. "Tell me what we have," she said.

The pilot looked at her with surprise and irritation. "Captain, this is a Naval matter. Passengers should return to their seats and strap in."

Lee was quiet for a long moment. She looked at the pilot and saw that he was a petty officer first class. The name tag read Ali.

"Petty Officer Ali, I am the senior Alliance officer on board this vessel," she said in a calm voice. "I also still hold a reserve commission as a Naval Lieutenant Commander." This last was technically true. When she had transferred to the Marine Corps, her former rank in the Navy had never actually been revoked. That would take an act of the Alliance senate, and she suspected they were too busy to bother with a lowly shuffle in the ranks. "I say again, *tell me what we have.*"

The petty officer reddened and then decided to comply. This Marine captain, whoever she was, wasn't going to take no for an answer. That she still had blood splattered on her body armor clinched the deal.

"Right, ma'am," he said. "We are on the very outskirts of the Dengbeh system. We were preparing to jump when this destroyer closed on us fast and has ordered us to heave to and prepare to be boarded."

"Let me see tactical," she said.

The pilot put it up on the screen. She looked for a moment and then reached over and pushed a few buttons. The screen zoomed in on the destroyer, then flipped to various views and magnification.

"It's a Victory class destroyer, older model, the Alliance phased these out decade ago," said Lee. "See the attitude jets on the port side toward the stern?"

"Yes, ma'am," responded the pilot.

"There's the same on the starboard side," she said. "See how they protrude from the rest of the vessel? They are vulnerable, the shield is weakest there. Take them out and the destroyer cannot maneuver. If they can't maneuver, they can't jump."

"You don't expect me to fire on a destroyer..." started the alarmed pilot.

"I expect you to do what I tell you," said Lee sharply. "Or are you planning to refuse a lawful order from a senior officer?"

The pilot looked over at his co-pilot, a young woman whose rank insignia showed she was a pilot in training. Her name tag read *Odessa*. The trainee shrugged.

The pilot said resignedly, "I acknowledge your authority and will take your orders. I am recording your assumption of command in the log."

"Glad to hear it," said Lee. "Now, show me navigation."

Lee looked at the display. "You will need to set a course at max speed to that jump point, she pointed to the screen.

"Show me weapons," she said. "Hmmm... Better than I'd hoped. You have twin Gatling guns and 50,000 rounds of armor penetrating 20-millimeter rounds. That will give you enough for two quick bursts."

"Propulsion," she ordered. After watching a moment, she said, "Thank God you have jump drive. Where is the next destination from the jump point?"

The pilot looked flustered and said, "I don't know. Lieutenant Danner was supposed to tell me once we were underway."

"Wonderful," said Lee. "Danner!" She shouted toward the passenger compartment. "I need you up here."

A moment later, Danner appeared, sleep still on his face. "What?"

"We're about to be boarded by a destroyer," said Lee. "I need to know the destination for the next jump."

To his credit, thought Lee, Danner didn't ask, 'What destroyer?' Instead, he said, "We're supposed to go to the TOI-700 system," said Danner. "But if we're running from an enemy destroyer, we can't go there. We have contingency jump locations. The best one if we're on the run is

TYC 7037. It has six stars all in orbit with each other, and dozens of planets. Easy to hide there if we have to."

"Let's' do that," said Lee. "Pilot, program that point for the jump destination, and now listen carefully to my instructions."

CHAPTER 22

A BOARDING OPERATION

Fifteen minutes later, Danner's shuttle was about to be boarded. The destroyer had identified itself as part of the Confederate Alliance Naval Militia Forces. Neither Lee nor Danner had ever heard of such an organization. Danner thought it must be something that had grown out the rebellion and succession.

Lee disagreed. It made no sense. The rebels were in no way able to stand up to the Alliance. It was true that some of the Navy had gone with the rebels after the split. But that was a relatively small force, and when the rebels left, they had taken modern vessels. This destroyer was, comparatively, a ramshackle piece of junk. It wouldn't last more than a few minutes against a modern Alliance warship.

It was much more likely that these were pirates. They were pretending to be a semi-legitimate authority as a way to take vessels who might otherwise fight or run.

Lee had insisted that the pirates be allowed to board.

The destroyer had sent its own mini-shuttle alongside the Alliance shuttle which had docked a few minutes ago. It was just large enough to hold its occupants and was designed only for ferrying personnel between vessels. Lee had used the time to strip off her armor. She sat in her seat, unbuckled, head and eyes down.

The bulkhead door opened and three men entered the passenger compartment of the shuttle. Lee watched them surreptitiously. They were dressed in odd, mismatched pieces of uniforms. They were armed with sidearms, and their hair was longish and uncombed. In general, they seemed unkempt.

Definitely not actual Naval personnel.

Danner met them as they entered. He was in uniform and had his sidearm on his hip.

"I am Lieutenant Danner of the Alliance Navy, commanding officer of this shuttle. Your entry is a violation of Alliance law and the intergalactic code section forty-five."

The man in front of the others, said, "Right, and your face is a violation of intergalactic code for preventing ugliness." As he said this, he drew his sidearm and whipped it across Danner's face, knocking him back in a spray of blood. The next man in line rushed forward and disarmed Danner as he lay stunned and bleeding.

Lee crossed her arms holding her shoulders tight and whimpered, "Oh God, no, please, please no."

The leader of the pirates came over to her. "Now, now, lass," he said. "No reason to cry. We'll fix you up nicely. No pretty girl has ever gone out the airlock on my watch, as long as she is compliant, anyway."

He reached down and cupped her chin in his palm, lifting her head. Lee said meekly, "You won't hurt me?"

"Well, not at first, anyway," he said with a sneer. "There *are* three of us, and more on the frigate. But you look tough, Asians always are."

Lee gently placed her hand over his hand on her chin. "I'm so glad," said Lee with a different voice. "But a little pain is OK."

In a single motion almost too fast to see, Lee rose from her seat and simultaneously twisted the man's wrist, causing an audible snap of bone and cartilage. He screamed in pain and outrage as Lee took his sidearm, pulled him close and placed the muzzle to his crotch.

She said, "This will only hurt a little, then." And pulled the trigger. The sound of the shot was very loud in the enclosed space, and the gore from the wound splattered the floor. As the man fell, Lee was moving.

The other two pirates hadn't moved. They were in shock. No passengers had ever fought back, and this tiny Asian woman moved like a she-devil, almost too fast to understand.

In a fraction of a second, she had hit the next pirate in the throat with an open-hand strike, he started to gasp for air that would not come, his hands raised to his throat.

Without pausing, she was past him onto the third man. He didn't have time to try and draw his weapon, so he just raised his fists in a protective measure to guard his face. Lee kicked him hard in the thigh, and he went down screaming in pain.

By now Zakany was tending to Danner, who was just starting to wake up from his injury. "The lieutenant will be fine," she said. "Mild concussion only." Zakany moved to the wounded pirate. "He also will live, no children though."

Lee quickly collected their weapons, and placed restraining ties on the hands of the two who were still conscious. She dragged them around on their backs, side by side, so they could both see her and each other.

Lee showed the cold face, the face her ancestors, the Mongols, had shown going into battle centuries ago. It showed nothing: not hate, nor anger, nor fear. The two men looked terrified.

"One chance only, gentleman," she said holstering her firearm and drawing a wicked-looking knife about six inches long. "I will speak; you will do. If not, I will leave you a gelding, like your fearless leader."

The two men both nodded frantically. One, she noted, had wet his pants.

"Which of you is senior?" When neither immediately replied, she knelt to the nearest man and methodically began undoing the belt at his waist.

"I am, I am!" shouted the man. "Please lady, whatever you want."

"You will contact the destroyer," she said, "Tell them you are bringing the shuttle aboard through the starboard docking bay."

"They won't do it," said the man. "The way we always do it is to pilot the vessels and have it towed back to our base."

"Tell them the hold has 12 tons of Rhodium," she said. Rhodium was the most valuable metal in the Galaxy. Twelve tons would make every man aboard the pirate ship wealthy.

Lee cut through his restraints. The man nodded and reached for the communicator on his wrist. Lee smiled and placed the knife at his crotch and said, "If you try to use a duress code, I'll know."

"I get it," the man said nodding furiously.

"Boss, this is Jones," said the man.

"Yeah, go ahead," came the reply.

"This ship's got twelves tons of Rhodium in the hold," said the man. "Freddie wants you to bring the shuttle into the starboard docking bay. He doesn't want to risk losing the load."

"Why aren't I talking to Freddie, then," came the reply from the Frigate.

"He's tied up with this cute Asian bitch," said the man. "She's a handful."

"Roger," said the man. "Come about, our shields are down."

"Pilot, *now*!" shouted Lee.

CHAPTER 23

ESCAPE

The occupants of the shuttle were thrown violently, as the pilot executed the maneuver Lee had ordered. He pivoted the Alliance shuttle so that it faced the destroyer, and then opened fire with both Gatling guns, targeting the protruding attitude jets of the pirate vessel, while at the same time accelerating toward the destroyer in what looked like a suicide charge.

The destroyer's starboard attitude jets exploded in a huge fireball as the armor piercing rounds ate into the unshielded flank. The pilot then skillfully shifted aim to the main navigation sensor pod on top of the frigate. It was a smaller target – a barely visible bump out of the top of the hull.

Lee was impressed. This shuttle pilot knew what he was doing. He first aimed at the body of the frigate and then walked the rounds upward until one finally smashed into the pod just as the shuttle flew past the frigate now heading in the opposite direction.

Lee, for her part, turned to Danner and said, "Get these three pirates into an escape pod and push them out."

Turning to Zakany, she asked, "Laura, is their fearless leader ready to travel?"

"Yes, Mei Ling," said Zakany. "He is patched up for now and sedated. The auto-doc in the pod will keep him alive. He will be OK as long as the destroyer recovers the pod in the next few hours."

Lee spoke to the two still-conscious pirates as Danner was dragging them to one of the escape pods. "Gentlemen, you have made a grave mistake by attacking us. One of you struck a Naval officer in command of an Alliance warship. That's a death sentence for all of you if you are caught. I suggest you ditch the destroyer and find another line of work. Do you understand?"

One of the men said, "Why are you letting us go?"

"I am letting you go and letting you live for one reason only," said Lee. "If you or your companions have prisoners you've taken in piracy, you will release them. From this moment on, they will be untouched. If you don't do as I have ordered, the *Asian bitch* will be back, next time with a real destroyer. I have your DNA; I can track you. This is your only warning."

After the pod had been ejected, Lee returned to the cockpit.

"How is our destroyer doing?" she asked.

"Ma'am," responded the pilot. "As you directed, the starboard stabilizer was destroyed, and we were able to put at least one round into the navigation pod."

"What is the destroyer doing now?" she asked.

"She has come about, facing our direction of travel," he responded.

"For what purpose?" she asked. "She can't pursue without two working stabilizers."

They both watched the tactical display.

"That," said the pilot grimly pointing to the screen.

"Shit," said Lee. "They've fired a missile."

The pilot hit a switch and a red flashing light came on, and the public address system sounded: "All passengers and crew, prepare for impact, missile inbound."

Back in the passenger compartment, Zakany and Danner were strapping into their seats.

Lee said to the pilot, "Show me the missile track."

On screen, Lee saw the predicted track of the missile.

"Time to impact?" she asked.

"Two minutes, ten seconds", he said.

"Can you evade it? she asked.

"Not forever, and if I try, we won't make the jump point."

"Shields?" she asked.

"If I put everything to the rear, the missile won't penetrate, but it doesn't need to. The kinetic force when it hits the shield, that alone will overwhelm the internal inertial dampers. We will be squashed from the acceleration."

"Can we outrun it or jump before it reaches us?" she asked.

"It depends, ma'am," he said. "That missile has a mark 25 seeker warhead. It's designed to seek its target and accelerate the last few hundred meters to achieve maximum penetration and to prevent a last-minute maneuver on the part of its target."

"So, what are you telling me?" said Lee.

"Just that there is no way to know for sure if the missile's final acceleration will reach us before we can jump to FTL. There's a variability in the programming, and the warhead's AI program allows it discretion in deciding the approach. It's possible that if the AI realizes we are about to jump away, it will accelerate sooner to try and catch up before we are out of its range."

"So, you're telling me, it's going to be close," said Lee.

"Yes, ma'am," said the pilot.

"Your recommendation?", asked Lee.

The pilot turned to look directly at her. "I recommend we make all haste to the jump point and take our chances. We can't outrun it in normal space, nor can we continue to evade it forever. If we stay, it will eventually destroy us."

"Any tricks you know that might give us an edge?" she asked.

The pilot made a wicked grin and said, "I'm a combat pilot, ma'am. We always have at least one trick up our sleeve. But I'll need some help."

CHAPTER 24

A CLOSE THING

In the approximately 90 seconds remaining before the missile impacted, Lee, Danner and the co-pilot worked frantically to make the alterations Petty Officer Ali had ordered.

"Ten seconds," announced Ali. "Best to strap in."

With only moments to spare, all three passengers and the co-pilot were strapped into their seats. Lee had chosen to occupy the jump seat behind the two pilots in the cockpit.

She watched on the screen as the missile closed with the shuttle.

A computer voice said, "Impact in *five, four, three, two...*"

At the very last moment before impact, the pilot executed the maneuver he planned. First, he used an explosive charge to eject the pirate's mini-shuttle that was still docked to the shuttle. The tiny vessel broke free of the shuttle, and a milli-second later the mini-shuttle exploded in a bright fireball. At that moment, the pilot turned hard toward the jump point, and then engaged the jump drive.

The shuttle shook violently like a rat in the jaws of a terrier. If they hadn't been strapped in, they would have been thrown violently against the bulkhead. Then the lights went out completely, and the interior of shuttle was uncharacteristically silent. The only sound heard was the automatically released oxygen masks bots that crawled around from behind each seat, up onto the faces of the unconscious passengers and crew, and began pumping lifesaving oxygen into their lungs.

Lee awoke in extreme discomfort. Her neck hurt terribly, and her insides felt like she had swallowed a basketball. She breathed in deep gulps of air before she realized she had an oxygen mask on. She looked at the reading on the display of the mask and saw that the air levels were recovering from an earlier dip and were at near normal levels.

Looking forward she saw that both pilot and co-pilot were unconscious, heads down, but breathing normally through their masks. She unstrapped and moved forward, crouching in front of the pilot's display. What she saw caused her to frown and then, as she cycled through the various programs, she sighed and said, "damn."

CHAPTER 25

FTL WITH AN UNINVITED GUEST

Lee checked on each of the occupants of the shuttle. The auto-doc of the shuttle's operating system provided immediate first aid and showed Lee the status of each. Bottom line, they were all OK. The shock of the impact and the immediate loss of cabin pressure and oxygen had caused an initial loss of consciousness.

But something else was at play and she was unable to figure out what it was. They should not still be asleep, and Lee searched the computer to figure it out. It was a mystery.

Eventually, after about an hour they began to wake up. The pilots woke up first. She checked each of them for concussion and general cognitive functioning. They were both groggy as they came around, but otherwise OK. She made sure they had water, and that the auto-doc had stabilized the critical biological systems.

Ali began to wake up first, looking panicked, and began to punch keys at his console.

"What the hell," he said with frustration.

"I've locked you out of your controls," said Lee steadily.

Ali turned his head and glared at her. "I'm the command pilot; you have no right..."

"Hang on, there," said Lee raising her hands in a placating gesture. "You've been unconscious for hours and I wanted to make sure you were sound before letting you start in with the controls. That *is* standard operating procedure out of the Alliance flight regulations."

Ali seemed to gather himself. "Yes, right. What is going on?"

"Let me unlock your system and we can try to figure it out together," said Lee. "But better check on your co-pilot first."

Cadet Odessa was stirring, and Ali tended to her until she was reasonably alert.

"OK," said Lee. "Let's get started. You two run through your checks and then when you have a grasp on the vessel and the tactical situation, let's discuss. I'm going to check on the passengers."

Lee walked back to the passenger compartment and found Zakany out of her seat and tending to Danner. Because of his injuries during the boarding operation, he had taken longer to regain consciousness. To Lee, he looked dazed and unfocused.

"His blood pressure is high, not dangerous," Zakany said. "But it is a sign that his metabolism is struggling. The mild concussion from the attack, followed by the jolt we received from the missile has taken a toll."

"What's going on," said Danner in a groggy voice.

"We're in jump status," said Lee. "But we took some damage going in. Something weird happened. I'm not sure what it is. I'm going up to the cockpit to discuss with the pilots. Best you stay here and heal up."

Danner tried to rise and said, "I'll come," but he immediately sat back down.

"Stay with me, Lieutenant," said Zakany. "Let's sort you out first."

95

Lee returned to the cockpit where both Ali and Odessa were running through checks and speaking in low tones. Lee sat in the jump seat, watching, but letting them work.

"There," said Ali to both of them. Lee looked at the screen and saw a blurry object suspended in place with numerous other smaller dots around it.

"What I am looking at?" said Lee.

"That, ma'am," said Ali, "is a video capture from our rear-facing camera."

"I don't get it," said Lee. "It looks entirely static. What's that large object in the center?"

"It's the missile warhead," said Ali, "Or at least part of it."

"How could that be?" said Lee. "Surely the missile already detonated. Otherwise, what was that rattling we took."

"Yes and no," said Ali. "You recall what we did. We split off from the pirate mini-shuttle at the very last moment and we set it to explode with magnesium charges so it would be the brightest, hottest thing in the sky. Our hope was that the missile's AI would track the mini-shuttle and not us."

"Is that not what happened?" asked Lee. "If the missile had hit us full on, we'd be dead."

"You're right, ma'am," said Ali. "Here's what we think happened. We thought it was a Mark 25 missile warhead. That's what the computer told us. The AI for a Mark 25 would have had to pick one target or the other. We gave it a choice and we hoped it would pick the mini-shuttle."

"And that's not what happened?" asked Lee.

"No, it isn't," said Ali. He pushed some buttons, and Lee saw on screen a slowed-down display video recording of the inbound missile moments before impact. The display showed the explosive separation of the mini-shuttle and then the warhead did something totally unexpected.

"It split in two!" said Lee.

"Yes, ma'am," responded Ali. "The warhead seeker must either have been modified, or else it's a more advanced model."

As they continued to watch, one half of the warhead went for the mini-shuttle, it hit and exploded with a fury. The other half tracked the Alliance shuttle. The blast and debris from the mini-shuttle overtook the Alliance shuttle, and then everything went dark.

"So, what does this mean?" asked Lee.

"What we felt was the explosive effect and debris from the mini-shuttle," responded Ali.

"And the other part of the warhead?" asked Lee.

"Came with us when we jumped," responded Ali.

"Why hasn't it blown?" asked Lee.

"It can't," responded Ali. "We're in jump space, which is not real space. The laws of physics are, in some way, suspended. The jump drive creates a shell around our vessel, and the warhead is inside it with us."

"So, it can't detonate?" asked Lee.

"No, it can't," said Ali. "Not until we drop out of warp. Then it will."

CHAPTER 26

A PROBLEM AND A TIME LIMIT

Lee stared again at the image of the warhead. "How close is it?" she asked.

"One hundred and twenty meters," answered the pilot.

"If it detonates when we come out of warp, what result?"

"At that distance, catastrophic, probably," he said.

"Why probably," she asked.

"Because I have no way of knowing its relative velocity to us," responded Ali. "I don't know if the warp field will alter that velocity. When we went to FTL, and trapped it with us in the warp field, it was moving at an oblique angle because of the way the warhead split off and because we were turning toward the jump point at the last second. Had we not gone to FTL, the warhead would not have made a direct impact with the shuttle. Instead, it would have detonated at its closest proximity, ensuring we got as much of the blast effect as was possible given the relative trajectories."

"Let me see the trajectory and detonation if we hadn't gone to FTL," ordered Lee.

Ali tapped some keys and a diagram of the trajectories appeared on the screen. Just as Ali had said, the missile's warhead trajectory would have passed the shuttle. The screen showed the point at which the warhead would have detonated and the blast radius moving outward until it overtook the shuttle.

"Now back it up and show me what really happened when we went to warp." said Lee.

This time the screen showed the same early trajectory, but when the shuttle went to warp, the screen showed a small conical shaped enclosure that Lee knew was the warp field. It wrapped around the shuttle and just barely captured the warhead. At that point, the image stabilized, and the warhead and the shuttle remained static compared to each other.

"Now show me what happens when we come out of warp, assuming the relative velocities resume the way they were when we began the jump," said Lee. "This time put a timer on screen."

Ali did as she ordered, and they saw on-screen the timer count as the projected path of the warhead and then detonate at the five-second mark.

"OK," said Lee. "That's our worst case. How much time until we drop out of warp at our destination?"

Ali glanced at the screen and said, "Five hours, twelve minutes."

"We have five hours and twelve minutes to figure out a plan and implement it to mitigate or eliminate the effects of that warhead," said Lee. "Let's get everyone involved. Can you leave the cockpit on autopilot?"

Ali looked at his co-pilot who nodded. He said, "Yes. Nothing to do while in jump status until we get closer to our dropout point."

CHAPTER 27

A CONUNDRUM

The five of them sat around a small circular table in the passenger compartment that was used for meals on longer trips. At Lee's request, Ali had used a hologram projection to explain the tactical situation to the group.

"Lieutenant Danner," said Lee. "You have an engineering degree. What do you think?"

"Hmmm...," said Danner. "To be honest, I don't have any grounding in jump mechanics. So let me tell you what I do know and see if that helps."

The group nodded and Danner continued. "The jump itself is predicated on the area surrounding the vessel being clear of all debris. That's one of the reasons we don't jump from just anywhere. It's why we couldn't jump sooner than we did, even though that would have been desirable to avoid the missile strike. In fact, the jump drive should not have allowed us to jump with the debris and the missile warhead within the warp envelope."

"I'm thinking the debris from the explosion of the mini-shuttle and warhead itself only entered the warp field after it was already forming," said Ali. "That's why the computer controlling the jump drive allowed us to jump. There is no evidence in the jump drive of a malfunction."

"Lieutenant Lee, you said I have an engineering degree," said Danner. "That's true, but I am not an engineer. I am not aware of any recorded event similar to our situation. There are cases where a vessel has gone into jump status under fire or in close proximity to an explosion. In all those cases the vessel never came out of warp. Since there is no way to search for a vessel in warp space, those vessels were presumed to be lost."

"Ali," said Lee. "Is the warp drive performing normally?"

Ali consulted his tablet. "Ummm... No. It's drawing more power than it should, and it's showing a greater than normal instability."

"What do you think is causing that?" asked Lee.

"All I can think of is that the computer assumes that the debris and the warhead are part of our vessel that's broken off," responded Ali. "It's likely struggling to preserve these items during the jump."

"Is that going to affect where we end up when we come out of warp?" asked Lee.

Ali was silent while thinking. Then he said, "The navigation program and jump drive computer have fail-safe mechanisms built in. If it looks like the jump drive is about to fail, the navigation program will automatically seek the best possible place to drop out of warp. Ideally, that would be near an inhabited planet."

"Can we influence that?" asked Lee.

Both Danner and Ali shook their heads. Ali responded, "No. The jump drive cannot be reprogrammed like that while we are in FTL. If the failsafe is triggered, the navigation system will take over the vessel. It's sort of like a reserve parachute. If the main one fails, the backup will automatically deploy. Same principle. It won't ask permission. It assumes saving lives is more important than a desirable destination. Again, as the Lieutenant said, this is very rare. I've heard of it, but I've never actually met anyone who survived a warp failure while in flight."

"How much time remains before we are scheduled to come out of FTL?" asked Danner.

"Four hours forty-five minutes," replied Ali. "I'll put the count-down clock up on the board so we can all see it."

"But we might not have that much time," said Lee. "If the warp drive becomes unstable, the navigation system's fail-safe will drop us out of warp when it sees fit."

Lee paused for a moment thinking. "Do we have EVA suits?" asked Lee.

"Yes, ma'am," answered Ali. "But..."

"But it's not clear it would be safe to use an EVA suit in warp space," said Danner. "The extra vehicular activity suit is designed for normal space. We already know that the laws of physics don't seem to function properly in warp. We have no way of knowing if the suit would even protect a person. It might be catastrophic."

"You're right," said Lee. "That would be a last resort."

She turned to Ali and said, "I suppose we have bots that can perform exterior repairs. Can we use one of them to try and push the warhead out of the warp field?"

Ali thought for a moment then said slowly, "Yes. That should work. I can get a repair bot outside the shuttle. If it were in normal space, I could program it to move to where the warhead is. What I don't know is whether the bot can maneuver in warp space. The warhead was stopped, frozen. I'm not sure if the bot can move in that environment."

"The next question is what we want the bot to do?" said Lee. "Push it out of the warp field? Redirect it? Disarm it?"

For a long moment, no one said anything. Then Danner said, "I don't think we should try to push it out of the warp field. That would probably destabilize the field enough to cause our computer to drop us out of warp. If the warhead was still inside the field, it would detonate as soon as we drop out."

Lee said, "I agree, that's too dangerous, unless we have no other option. What else?"

"We could try to redirect it so that the blast, when it happens, will not hit us head on," said Danner.

"Petty Officer Ali," said Lee. "What do you think?

"Ma'am," he said. "I can work out an angle that would result in a trajectory that would be least harmful to the shuttle. But I cannot guarantee the warhead will even be able to change direction. Angular momentum is conserved. Even if the warhead's AI wanted to change direction, it couldn't do it instantaneously."

"Anything else?" asked Lee.

Zakany, who had up until this point been silent, said, "Why not disarm the warhead so it can't detonate?"

Lee looked at Danner and then Ali with a questioning look on her face.

Danner said, "We could do that, but a bot could not. To clarify, the bots we have onboard could not disarm the warhead. They are not sophisticated enough."

"So, what you're saying," said Lee. "Is that if we need to disarm the warhead, a person will have to do it, not a bot."

Danner looked at Ali for confirmation, who nodded agreement. "Yes," said Danner.

CHAPTER 28

A HAIL MARY PLAY

Lee stood in the airlock in an EVA suit. "Stand by for final check," said Ali through her coms system. Then a moment later, "Beginning decompression of the airlock."

Lee had insisted that she be in full EVA ready to deploy in case the bot failed to function outside the shuttle in the work-space environment.

When Danner and Zakany had objected, she said, "Thank you all for your input. However, I am the senior Alliance officer present, I have taken command of the shuttle, and the decision is mine. I will suit up and be prepared to go EVA if the bot cannot redirect the warhead."

She turned to Ali and Danner. "You two will need to give me a crash course in disarming the Mark 25 warhead. I have been given general training in disarming explosives, but I will need the specifics for this one which is unfamiliar to me."

That was 30 minutes ago. Now she was in the airlock, waiting to see what the bot would do. She would be able to watch its progress on her heads-up display in the face plate of the EVA helmet.

"The bot has been released from its bracket," said Ali. For a long moment, the spider-shaped bot just sat there motionless.

"Is it responding?" asked Lee.

"The bot is acknowledging the commands I've sent," said Ali. "But it's not moving and has sent a quirky message that I don't understand."

"What does it say?" asked Lee.

"It's in computer code," replied Ali, "but it translates as roughly, 'Where am I?'" There was a pause and then Ali said, "I've re-transmitted instructions, along with coordinates as to its location. It hasn't responded yet; I can see from its readout that it is having trouble responding to the commands."

"How much time remains before we drop out of warp?" asked Lee.

"Two hours, 17 minutes," said Ali.

"Let's give it another minute," said Lee. "Then I'll go."

After 30 seconds, Ali said, "The bot is asking what to do about the others in the warp space?"

All were quiet for a moment and Lee said, "Any idea what it's talking about?"

"It's confusing," said Ali. "The bot is programmed to be exceptionally careful about humans during EVA because of the danger of interfering with them when they are vulnerable outside a vessel." He paused than said, "I just don't know. I can't see anything on our rear-facing sensors or camera."

"Could the bot be mistaking the debris from the explosion as humans in EVA suits?" asked Lee.

"Not likely," said Ali. "That would be a major fault. The bot's system has a calibration effect for that type of error."

"Suggestions, anyone?" said Lee.

Danner piped in, "Petty Officer Ali, can you override the bot and force it to go perform the task?

"Yes, up to a point," responded the pilot. "It will not do anything that could result in a high probability of serious injury or death to a human being. I am sending new instructions now." A pause then, "There, I've told it Captain Lee is the only human it could encounter during this mission. Let's see what it will do."

On the camera they all watched the bot unfold its spider-like legs. It's sensor eyes scanned the area in front of it, but it didn't move.

"Now it's asking if the others are hostile, and is it free to defend itself if attacked," said Ali.

"It's obviously suffering some kind of fault," said Lee. "Given the lack of time remaining, I think you should tell it whatever it needs to hear to get it going. If it doesn't do it, I'll go out."

"Will do," said Ali. "I just told it you are a friend, any others to be encountered may be considered hostile. Since there's nothing else out there, it shouldn't be a factor."

The bot immediately sprang out from its niche on the back of the shuttle and moved rapidly toward the warhead.

Lee said, "Well at least that worked..."

Just then the bot spun to its right and fired a short burst from its onboard laser.

"It's reporting hostile engagement and is firing its laser in self-defense," said Ali in a steady voice.

"Any harm done?" asked Lee.

"Its laser is designed for salvage work, not for combat," responded Ali. "It has a limited range, but even so, it has disrupted the warp field. Give me a moment. There, I've shut the bot down to prevent it from doing any more harm. And I've been able to stabilize the warp field by shifting energy to the damaged portion."

"I'm up, then," said Lee. She pushed the button that activated the opening of the airlock hatch.

CHAPTER 29

UNWELCOME THOUGHTS

Lee stepped out of the airlock, intending to wait a moment for the EVA suit to adjust to the new environment and confirm the various systems were functioning. She didn't get the chance.

Lee was immediately hit with an overwhelming sense of disorientation and nausea. She heard a rushing sound in her ears, and her vision was... Well, there really wasn't a word for what she was seeing.

"Whoa," she said to herself, but the exclamation was transmitted to her colleagues on the ship.

"Mei Ling," said Laura Zakany. "Are you OK? We can't read your vital signs."

For a long moment there was no reply, and then Lee said. "This is different... Vertigo, visual distortions. Wait..."

Lee couldn't believe what she was seeing. Faces, moving quickly past her field of vision, sounds – no, there were voices — indistinct and then too loud. She closed her eyes and breathed deeply and steadily. Whatever this was, she needed to deal with it.

"I'm having some sort of auditory and visual hallucinations," said Lee. "I can't see the target... What was it?"

"The target is the warhead," said Danner calmly. "I'm going to put a beacon on it. Your suit should be able to go to it automatically. Do you see the beacon?"

Lee searched her field of vision, and indeed she did see a blue blinking light ahead, but the faces and voices were still there.

"I can see it," said Lee in a distant weak voice. "I'm setting the EVA suit computer on auto retrieval."

Lee was concentrating on her task. What was it? Oh, yes. Disarm the warhead. She was tired. The effort to shut out the voices and images was difficult. She realized she was in danger of going into shock.

"Laura," said Lee to Zakany. "Can you read my vitals?"

"No, Mei Ling," said Zakany. "There is some sort of interference. Tell me your symptoms."

"Fatigue, nausea, disorientation, some memory issues, also I feel cold," said Lee. "Not well."

"Signs of impending shock," said Zakany in a calm voice. "I can't control the EVA suit's auto-doc from here. You can do it. Cycle through the suit's menu and select *medical*, then *shock*, then *epinephrine injection*. Once you've done that, the suit will give you a shot. It should help the shock symptoms."

As if in a dream where she was caught in thick molasses, Lee did as instructed, and soon she felt the little prick in her arm that told her she had been given a shot. In a minute, she started to feel a little bit better, warmer and less nausea.

"Got it," she said. "Feeling better." As Lee moved toward the warhead, the faces seemed to notice her and were clamoring outside the suit, trying to get her attention, apparently. "These bastards are persistent, though."

"What was that?" said Danner.

"Just these hallucinations," said Lee. "Mission first, I'll tell you about this later."

"You should be about a meter from the warhead now," said Ali. "Disengage the auto-retrieval and take it in manually."

Lee did so gradually. The warhead was about three feet long and a foot wide, tapered to a point at the nose and flat at the propulsion end.

Lee extracted the tools she had brought from a small pouch on the belly of the EVA suit.

"Walk me through the steps, please," said Lee.

"First," said Ali. "Touch the disarming probe element to the warhead's pointed end, and then wait for a data connect link."

The probe was a flexible rod about a foot long. Lee reached out and touched the warhead and...

CHAPTER 30

A DIFFERENT PLACE ALTOGETHER

Lee woke slowly. She knew she was injured, on her back. She opened her eyes and looked up at a yellow sky. She could hear the soft humming of insects near her, she could smell the earthy scent of vegetation.

A face appeared above her, human, but different. Eyes large, bright blue, forehead large, skin dark as if tanned by long years in the sun.

The face said, "There you are."

Then Lee was back in the shuttle with Dr. Zakany over her, working, saying something to Danner.

"What happened?" asked Lee.

"After you disarmed the warhead, you passed out," said Danner. "When your EVA suit detected your vital signs dropping, it brought you to the air lock."

"Mei Ling," said Zakany. "You were flatlined when we got you back. Somehow, you seem OK now."

Lee said, "I don't remember disarming the warhead. Are you sure I did it?"

Danner and Zakany looked at each other.

"Ali walked you through it," said Danner. "You were disoriented, and some steps had to be redone, but you did get through it. We now have your EVA suit video cam recording if you want to see it."

"I had a vision," said Lee. "I was somewhere else, on a planet. I saw a face. It was *very* real."

Lee sat up and looked around. "How much time until we drop out of warp?"

"Seven minutes," said Ali over internal coms from the cockpit. "Best to strap in."

As Danner, Zakany and Ali strapped into their seats, Ali said, "Just so you know, we *are* dropping out early because of the damage done to the warp field."

"Where will we be?" asked Lee.

"We'll still be in the TYC 7037 system, but I have no way of knowing where. The system is huge, multiple light years in circumference. It has six stars and dozens of planets."

"What's your estimate on the damage we'll take when we come out of warp?" asked Lee.

"There is no way to know for sure if the disarm was successful," responded Ali. "You definitely did all the steps. But we don't know if being in warp effects the result. Also, you passed out before we could run any type of verification tests."

"Any other good news?" asked Lee.

"Yes," replied Ali. "We still have the remaining debris from the mini-shuttle blast inside the warp field with us. There really *is* no way to know how much of that will hit the shuttle once we come out. Too many variables to be able to calculate. We'll certainly get some of it. Bottom line: hang on."

"What's the contingency if the shuttle breaks up?" asked Lee.

"Your seats are self-contained ejection pods," said Ali. "If the shuttle becomes untenable, it will eject you. If we are in — or close to —

an atmosphere, it will try to find a suitable place to set you down. If not, it has a short-term life support capability in vacuum. Hours, not days."

As the three of them strapped in, the seat automatically began enclosing them with protective restraints starting at their feet. Before the last of these closed over their faces, Lee said to Danner, "What's the mission?"

"What?" said Danner.

"You plucked me out of command for a reason," said Lee. "I still don't know what that reason is."

Danner looked confused a for a moment. "Didn't General Motubu tell you?

Lee's look told him *no*.

"It's Chambers," said Danner.

Then their world came apart.

CHAPTER 31

A VERY HARD LANDING

Several things happened in rapid succession. The shuttle dropped out of warp, the blast and debris from the explosion of the mini-shuttle hit the vessel and flipped it bow over stern, and the shuttle entered the atmosphere of a planet.

Inside the shuttle, the effect was catastrophic. The internal dampers lessened the deadly effect of rapid deceleration, which otherwise would have squashed them to jelly. As it was, the G-forces were extreme. The seat restraints saved them from broken necks.

Lee watched in horror as the shuttle broke apart around them, the outer hull ripped away in pieces. Through the gaping holes she could see a star field and planet whipping by in succession. Clearly the shuttle, what remained of it, was tumbling out of control.

Then she heard the computer's calm voice say, "Stand by for ejection." A second later, she felt the explosion of the ejection seat. It shot her backward and out of the shuttle. Once again, she was in free fall.

At first, the ejection seat was spinning wildly so that in Lee's field of vision, the sky and the ground were a blur of undifferentiated motion.

The seat itself had rudimentary attitude controls and Lee used them to manually steady the seat so that she was facing the planet's surface.

She was unfamiliar with the design of the seat, so she tried talking to it. "Computer, are you there?"

"Yes, Captain Lee," came a somewhat female digital voice. "How can I help you?"

"Please tell me my location and status," said Lee.

"You are sixteen thousand meters above planet four, of star sigma seven in the TYC 7037 system," said the computer. "You are in free fall," it added helpfully.

"Show me the planet's stats on my HUD," said Lee.

What she saw scroll across her Heads-Up Display encouraged her: a warm planet, average temperature 24 degrees Celsius, breathable air, fairly mild seasons, early summer now. She was falling toward a landmass. Better than the ocean, she thought.

"Show me the location and status of the other members of the shuttle," Lee ordered.

After a pause, the voice came back, "I'm sorry, Captain. I am only the computer for your ejection seat. I have no data on the location of other occupants."

"Are there any retrieval beacons sounding?" she asked.

A pause, then, "Not at this time, Captain, other than yours," came the voice. "However, there is a great deal of electromagnetic interference in the atmosphere. Given the speed of the entry, it is likely that any other escape pods that were ejected before disintegration might well be out of range or over the horizon from you."

Lee noticed the terrain below her getting closer and wondered how the ejection seat was going to make a soft landing.

"How does the landing work with this seat?" asked Lee.

"This model ejection seat has twin parachutes to reduce the rate of descent, plus a single backup chute in case of main chute failure. On final approach, the model seat has deceleration jets to slow your speed to expected survival levels."

Lee noticed a red blinking light in the corner of her heads-up display. "Computer, what's this red light mean?"

"Captain Lee," said the calm voice of the computer. "The red light indicates critical failure of the main and backup parachutes."

"Great, thanks for telling me," said Lee sarcastically.

"You're welcome," came the polite reply.

"What's wrong with the parachutes?" asked Lee. "Can they be made to deploy?"

"No, Captain Lee," came the calm reply. "The housing that contains the parachutes was torn away upon ejection."

"What about the deceleration jets?" asked Lee.

"That feature is intact," came the reply.

"I *meant*, can I use the jets to slow my descent so I can survive the impact?" Lee said with a forced calm that she did not feel. She saw the ground below rushing toward her.

"Negative, Captain," said the computer. "You are currently traveling at 60 meters per second, which is terminal velocity. Use of the jets will slow that to 55 meters per second, which is not a survivable scenario."

"Options?" asked Lee.

"You can fly," said the computer in its flat voice.

"What?!" said Lee.

"I have detected that you are wearing a FW 287 Special Operations Battle Armor, Wing-Suit equipped," said the voice. "You can fly."

Of course, thought Lee. She had been wearing her battle armor when she ejected.

"Got it," said Lee. "Compute optimal altitude for disengagement from seat to commence wing-suit descent."

"You are past optimal altitude," said the only calm voice. "Approaching minimum survivable altitude in five, four, three…"

"Computer," said Lee. "Initiate emergency disengagement from this seat."

"Yes, Captain," said the ever-calm voice. "Goodbye."

Instantly, Lee was blown away from the seat by an explosive charge. Because she had been facing the ground, the seat shot her downward like a cannonball toward the surface.

She was way too close to the ground, a mere few hundred meters. She deployed the wing suit function on the suit, tried for the shallowest glide path possible, and felt the extreme stress on the wings of the suit as it rapidly slowed her descent. Then she tried to pick a path to relative safety.

Damn, thought Lee. The terrain below was mountainous and forested. The only relatively open space was a meadow about three miles from her current location. Lee wasn't sure she could make it. She was just traveling too fast and running out of altitude. She adjusted her flight profile and raced toward the meadow. At the speed she was moving, anything other than the meadow was going to be fatal.

She was in an aviator's classic dilemma. If she bled off airspeed, she would lose altitude and crash into the forest. If she maintained speed, she would make the safety of the meadow, but would be traveling too fast to make a safe landing. Beyond the meadow, the terrain dropped off precipitously; it was basically a canyon. That wouldn't do either.

No, she would have to go for the meadow and take the hard landing. But even that wasn't a sure thing, the air was thinner than she was used to, and she continued to lose altitude faster than she needed. In the last few seconds, she knew she wasn't going to make it.

And then it happened. She hit an air pocket and dropped a few meters and slammed into a treetop just shy of the meadow. And then nothing.

CHAPTER 32

THERE YOU ARE

A face appeared above her, male, the wrinkled face of an old man, but with youthful movements and expressions. He was human, but different. Asian features, but large bright blue eyes, unlike any Asian she knew. The man's forehead was pronounced and the skin was dark as if tanned by long years in the sun.

The face said, "There you are."

It was the same man from the vision in the warp field she had seen a few hours before, but the man looked older than he had in her vision, somehow more wrinkled. Lee struggled to speak, but was only able to croak out the word, "*Before...*"

The man turned to speak with someone outside her view. "It's a woman, a warrior, she has her enemy's blood on her armor, she is the one I told you about."

Another voice said, "And she fell from the sky? Are you sure?"

"Yes," replied the first man. "I told you; I saw it. She was flying in that suit and then she hit the tree." He pointed back over her head. "Do you see the branches are broken?"

"She's hurt," said the man outside her view. "Can she be repaired?"

"I'll ask," said the first. "Hello, I am Guyuk. Who are you?"

Lee was struggling to understand what was happening. *Guyuk?* That was a Mongolian surname. Lee realized the two men were speaking a dialect of Mongolian, a language that she had spoken at home as a child with her grandmother.

Lee struggled to sit up, then got to her knees, then stood up facing the two men. She was dizzy, maybe another concussion, she thought. She had scratches on her face and an open wound to her right thigh.

The two men were short in stature, a few inches shorter than her 5-foot 2-inch frame. Both were completely bald or had shaved their heads. They were wearing purple and white robes that had sleeves to their elbows, with the leggings ending just below the knees, tied with a cloth sash at the waist. They wore rugged-looking shoes.

And they looked *solid*, with ropy arm muscles and wide, strong calves showing below the robe. *Monks, perhaps*, thought Lee. *But here?* She thought to herself, *best to let this play out.*

"Zolgokh," she said as held out both arms, elbows at her ribs, palms up in a hugging gesture. This was the traditional greeting for Mongolians.

The first man, the one who had spoken to her, smiled broadly, came forward and clasped her forearms with his under hers, his palms upward cupping her elbows.

"Elder monk," she said in what she hoped was understandable Mongolian, "It is I who should defer to you."

The man had clasped her with his arms below hers; a sign of social subordination. As the younger, she should have shown the traditional deference by placing her arms beneath his. But he had grasped her so quickly she couldn't have stopped him without making an awkward gesture.

"No!" He spoke sharply. "You are the warrior goddess of prophecy."

118

"Sir, kind monk," she said. "I am not a goddess. I am Mei Ling Lee. I am mortal, an Alliance officer. My ship has crashed..."

"Yes!" He said with eagerness. "We saw it. A streak of light, and massive boom," Lee realized he must have heard the sonic boom as the shuttle entered the atmosphere and broke up.

"It was foretold," he continued. "And here you are."

"I saw you," Lee said hesitantly. "In a vision."

"Yes, exactly!" said Guyuk excitedly. "Many years ago, I saw you in this meadow in a dream."

"For me," said Lee. "It was only a few hours ago, not many years."

"And such is the way of prophecy," he relied. "It is not according to our way of experience. Time belongs to the gods."

"When I had the vision, so many years ago," said Guyuk with reverence, "I told the Abbot, and he wrote it down. He said it foretold the rebirth of Daichi Tengri."

Lee knew enough about Mongolian mythology to know that Daichi Tengri was the Goddess of War. But in the modern day, few Mongols believed such myths. But this wasn't Earth, certainly not Mongolia.

"Sir, *I am not Tengri*; not a goddess. I promise you. I am captain of Marines; no more than that."

Guyuk shrugged and said, "It is of no matter. Those who are incarnate, often do not know their true self when born again into life."

She pointed to a wound on her thigh. "As you can see, I am wounded, bleeding, and not a goddess."

And as if timed to prove her point, she promptly fainted.

CHAPTER 33

A SAFE PLACE

Lee woke. She didn't know where she was, so she listened first. Muffled sounds, a fire burning in a hearth, soft voices. She breathed in deeply, the smell of stone, candle wax, wood burning, and of incense.

She opened her eyes. She was in a darkened room, partially illuminated by a lit fireplace in one corner, and candles on a nearby heavy wooden table. She was on a simple cot with a course blanket of wool pulled up to her shoulders.

She sat up. She was no longer in her battle armor, but was wearing a simple cotton shirt and pants. Her leg wound had been bandaged.

She had a moment of panic when she realized she didn't have her weapons, but then she saw all her kit, including the armor, firearms and knives neatly folded and stacked on a nearby table.

She sat up, looked around to make sure she knew the outlines of the room, and said, "Hello."

A moment later, two women entered the room through a curtained doorway. Each was clothed in traditional Mongolian dress for women that included a pearl head dressing, silver earrings, and long, flowing gowns of dark colors, reaching to the ankles.

One was older, maybe 50, hard to tell. She walked forward toward Lee in a confident and relaxed manner, looking Lee in the eyes. The other was younger, maybe late teens early twenties. She kept her gaze down and stayed a step behind.

"Good morning, Madam Warrior," said the woman in clearly enunciated but accented Mandarin. "I am Sarnai, your hostess and guide while you are here."

"You speak Mandarin," said Lee.

"Yes, ma'am, poorly, I'm sure. We have had no native Mandarin speakers here for a long time. It is a second language for those who wish to remember."

"And you don't address me as Daichi Tengri," said Lee. "I thank you for that."

Sarnai smiled and said, "I spoke with Guyuk, and I know he is convinced of your divinity. Clearly you are human, flesh and blood. Whether or not you are the incarnate of a god, or not, you have indicated your wish to not be addressed as such. It would be impolite to contradict your claim of identity."

"Why address me in Mandarin?" asked Lee. "Is my Mongolian so poor?"

"Honored guest," said the woman with a slight bow. "Your Mongolian *is* poor, no offense intended. And you are not Mongolian. You have not the features, and *Mei Ling* is a Chinese name, not Mongolian."

"I am descended from Khutulun, cousin to Kublai Khan," said Lee. "My family remained in China after the conquest in 1279. Over the years, we melded into the Chinese society."

The woman looked confused, opened her mouth to speak, shut it, and blushed red at her own awkwardness.

The other woman, the younger one, snapped her eyes up at Lee and blurted out in Mongolian, "A *Khan* conquered *the Celestial Empire?!*"

"Hush, Child," barked the elder woman. She looked at Lee and said, "We both exceed our place, I'm afraid. May I take you to freshen up and then for some food? Our Abbot wishes to formally welcome you."

CHAPTER 34

A BATH, A MEAL, A WARM WELCOME — AND AN AWKWARD CONVERSATION

Lee was led by the women into a bath area. There were a series of tubs of steaming water, with hand cloths and hanging towels nearby. She gratefully stripped down, removed her leg bandage and saw the wound was not deep, and then submerged herself in the hot, soapy water.

Lee had not taken a bath or shower in days, and she still had sweat, blood, dirt and leaves encrusted in her skin and hair. She used the rough washcloths to scrub her skin until it hurt. Lee had been in life-or-death circumstances several times in the past 24 hours. It was glorious to be clean and alive.

Afterward, she toweled off and found that her guides had left for her traditional Mongolian clothes, which she gratefully donned. There was also a traditional pair of sandals, which felt wonderful on her feet.

As she was finishing, her guides came back and led her through the entrance into a large hall, with great beams of wood used for the walls and ceiling. On the far side of the hall, she saw a table set with food, and beyond a huge pit with a roaring fire. The food smelled wonderful. Lee realized she hadn't eaten any food since before the mission against the slavers. She was famished.

As she approached the table, she saw three people who stood as she drew near. In the center was an older gentleman dressed as a monk in simple robes made of a course material, the Abbot she assumed. With him were two others. On the Abbot's left was a woman, middle-aged, straight-backed, severe countenance, also dressed simply.

On the Abbot's right was a man, younger, maybe 35, not dressed as a monk. He had dark, collar-length hair, and a rough but trimmed beard. He was tall and powerfully built. He looked at her directly as if assessing her. *Almost certainly a warrior,* thought Lee.

As she approached the table, Sarnai announced in Mongolian in a formal voice: "Lord Abbot, may I present Human Alliance Marine Captain Mei Ling Lee. Captain Lee, may I present Senior Abbot Altan of the Ongii Monastery."

At this, the elder man bowed and said, "May I also present Abbess Terbish and Noker Aruban." The Abbot spoke in Mongolian in a surprisingly deep and strong voice.

Lee bowed low. The titles Abbot and Abbess were self-explanatory. *Noker* she knew was a title that typically meant *general*, but that rank could be used for any military leader.

"Captain Lee, we are honored at your presence. I trust your road was true," said the Abbot.

Lee understood this was a formalized greeting for a stranger, and she searched her memory for the correct response her grandmother would have made.

"Lord Abbot, Abbess, General," said Lee. "My road is true, my eyes are clear, and my heart is filled with gratitude."

Lee could see immediately that her response was correct, as the group seemed to relax and, from the corner of her eye, she saw that Sarnai smiled just a little bit, then her face quickly reverted to a formal mask.

The Abbot, made a gesture to the open seat, and said, "Please, Captain, sit."

Lee took the seat indicated while Sarnai set beside her. Immediately, monks brought finger food of small slices of seasoned bread and some kind of cooled drink poured into a metal goblet. Lee had to consciously force herself to wait until the host began to eat first, as courtesy demanded.

But to her surprise, the host, the two guests and Sarnai waited patiently for her to begin.

Perhaps from hunger or fatigue, Lee broke out laughing and said, "Lord Abbot, I think if you are waiting for a goddess to start eating, we will all go hungry tonight."

The Abbot stared at her, eyes wide in astonishment, then he also broke out in a deep laughter, the other guests looked at him in amazement, and then they also started to laugh. The Abbot finally stopped and wiped his eyes with a cloth napkin.

"Captain, you have brought honor and good cheer to this table," said the Abbot. "May I suggest we all begin to eat now, as the demands of courtesy have been satisfied."

Lee smiled and dug in with honest joy. The drink turned out to be made from some sort of mixed fruit, and she was glad to note it was not alcoholic as she did not enjoy alcohol. The food was amazing. Lee had to be careful not to overeat, as her hunger drove her to desire all that was before her.

The main meal consisted of stewed vegetables, more bread and, what surprised her, some sort of meat. *Not Buddhists that's for sure*, thought Lee. Afterwards, they were served some kind of strong tea and a sugary biscuit. Again, Lee found it delicious.

Lee waited calmly for the conversation that must now come.

"So, Captain Lee," started the Abbot. "We have many questions and I'm sure you do as well. Would you like to go first?"

Lee took a deep breath, and said, "Lord Abbot, my starship broke up as we entered the atmosphere. With me were four colleagues. I humbly ask your assistance in locating them and providing them aid if they survived."

The Abbot looked toward Noker Aruban and nodded his head as if to say Aruban should respond.

"Captain Lee," he started. "When we saw the trail of smoke and fire, and heard the thunder which came with it, we knew something had entered the atmosphere. That is a rare, but not unheard-of, occurrence. When the monastery reported your arrival, coming from the sky as you did, we sent out search parties to seek any survivors. It may be days or even weeks before we have news. The area we need to search is rough terrain and we do not control all of it. There are other..." He hesitated as if searching for the right word, "... *tribes*. And we will need their assistance and permission to enter their lands."

Lee said, "Thank you, general. I would be grateful to speak of the search efforts in more detail at your leisure."

"I would welcome that, Captain." Aruban paused and then said, "Captain, you very politely addressed me as *general*. My rank is actually not so high. In my case, *Noker* simply means leader. My rank is commandant, and I am the commander of the tribal constabulary. We are responsible for local defense and public order."

As there didn't seem to be a response from her required by the exchange, Lee simply nodded.

Lee turned back to the Abbot. "Honored Abbot, I am at a loss to understand how a Mongolian monastery is located on this planet so far from our home world."

The Abbot looked thoughtful and glanced at his two companions before answering. "Our transition to this place took occurred outside of living memory. We are curious about what you can tell us of our home."

"Have you not had any contact with Earth since you moved here?" Lee asked with surprise.

"None," said the Abbot.

"Was it the Alliance that brought you here?" she asked. "I was previously in the Navy, and I don't recall any settlements recorded in this sector."

When the Abbot shrugged, she asked, "How long ago was the migration?"

"We were brought here in the Year of the Goat, 3896," said the Abbot.

Lee stared, at first with her mouth open, then she shut it quickly. The date was impossible.

"Lord Abbot," Lee began cautiously. "I am confused, forgive me. May I ask what major events were happening on Earth at the time you came here?"

"Why of course, Khan Temüjin had just united the majority of the clans."

Lee said nothing, the dates matched. The Year of the Goat 3896 was indeed the year 1200 on the modern calendar. And it was the year Temüjin, later known as Genghis, united most of the Mongolian clans.

The Abbot said politely, "You told Sarnai that you were descended from Kublai Khan. I am afraid that is not a name we know. Can you tell me of him?

"Lord Abbot," said Lee. "These dates have me spinning. So let me answer as best I can. Kublai Khan was the grandson of Temujin Khan. Temujin completed the unifications of the Mongols and conquered great vast lands to the East, North and South. His grandson, Kublai Khan, eventually conquered the Song Dynasty in 1279 on the Gregorian calendar.

The other three drew a collective breath. Clearly this was news to them.

The warrior Aruban spoke up. "Captain Lee, I would never seek to question the honor of a guest, but I must ask for clarification. You are

127

saying the Mongolian Khan conquered the *Song Dynasty*? *China*, the *Celestial Empire*? This seems an impossible feat. Surely, you mean Kublai won some sort of a battle against the Song, not that he conquered all of China?"

"Honored warrior, I know my Mongolian speech is poor," said Lee. "Do you speak Mandarin? It will be easier for me to explain."

"Yes, of course," he responded in Mandarin.

Lee switched to Mandarin, "Kublai Khan, the great Khan of the Mongols, defeated the forces of Emperor of the Southern Song, at the battle of Yamen. The Song's Child Emperor, Zhao Xian, was deposed and sent into a monastery. The Mongols established the Yuan Dynasty which ruled China for 90 years before falling to the Ming, who were ethnically Han."

"And now?" asked Aruban. "How do the Mongols fair?"

"Mongolia is its own nation, with Russia to the North and China to the South, East and West. It covers the territory from the Gobi Desert to the Altai Mountains. Part of the Mongolian population lives within the borders of Russia; it is called Inner Mongolia. It has a mix of cultures and nationalities: Mongolian, Chinese — Han and Manchu."

The group was silent for long moments, and Lee could tell each was in their own thoughts about a homeland they had never seen, and to them was almost a mythical place. Lee knew to let them have this time of contemplation.

At length, the Abbot looked at Lee and said, "Captain Lee, you must have questions for us. Please ask what you will."

Lee took a deep breath to focus her thoughts and then said, "Lord Abbot, you said your community has been here since the Year of the Goat 3896 on the Chinese calendar. By my calculations, today's date is 5119."

"Yes, that is correct," said the Abbot with a raised eyebrow. "Is there a discrepancy?"

"No, sir," said Lee. "We agree on the current date." She paused, and said slowly, "but that would mean you've been here for over a thousand years."

The Abbot nodded and said, "Yes, 1,223 years." He paused and looked at her inquisitively. "You seem troubled by this; why?"

"Lord Abbot," she replied. "Much happened after you left. There was a time of great turbulence. Weapons of incredible destruction were used, not just on armies, but on the civilian populations. The order of nations collapsed into chaos that lasted for a century. Finally, the Alliance was formed, a world federation of sorts."

"I don't see why any of that would be relevant," said the Abbot in a polite voice.

"One of the things that allowed the Alliance to form was the discovery of faster-than-light travel," said Lee. "It allowed people who might otherwise have been in conflict with each other over resources to leave Earth and establish new homes of their own making. But that process did not begin until about 300 years ago." Lee shrugged and raised her hands in a questioning gesture. "This planet, where we are right now, is 1,900 light years from Earth. How did your ancestors come here 900 years before the discovery of interstellar travel?"

The Abbot looked surprised and said, "Why, *you* brought us here, of course."

CHAPTER 35

THE LOST TRIBE

Lee breathed deep the mountain air as she hiked the well-worn path. God, this was lovely. Early morning, sunrise, the smell of the trees, the cool air, the feel of sweat on her skin. Lee was thinking she had spent too much of her life in space in cramped, stuffy vessels. This was a glorious reprieve.

Sarnai, her guide and guard, stayed about 30 meters behind her. The Abbot had insisted Lee be escorted if she were to leave the grounds of the Monastery. Lee made clear to Sarnai that she would find her own way. If Sarnai wanted to come, that was her business, but Lee wanted to be alone.

As she walked, she began a meditative breathing routine designed to calm her spirit. After only a few moments, she felt the stress begin to leave her. She needed to be clear-headed to take stock of what was said last night.

The Abbot had claimed that in the Year of the Goat 3896, over 1,200 years ago, a warrior had come out of the sky with a huge star ship. She didn't say her name, but later she was believed to be Daichi Tengri, a Mongol goddess of war.

The goddess told the clan leaders that she had sure knowledge that the entire clan would be destroyed unless they came with her. She said she would lead them to another place, a world, where they could restart without fear from enemies.

The enemies she spoke of were the soon-to-be-united armies of Temujin Khan. Temujin was gaining power, welcoming those who would submit to him, and utterly destroying those who would not. The tribe, *the Dukha,* resided in a mountainous region on the border of Mongolia and Russia, far from the Steppes where the wandering tribes of mounted Mongols were now gaining power.

"I take it the Dukha refused to join Temujin," Lee had said.

"Yes," responded the Abbot. "Our ancestors had no connection to the nomadic tribes of the plains. We were — and are — farmers and reindeer herders."

"That's it?" asked Lee. "Surely, they knew refusing the khan would be a death sentence for the tribe. Why not submit?"

"There were other reasons. Temujin claimed to be the incarnation of the Sky God Tengri. Our ancestors were Tengrists. They knew Temujin was a false god, a fake. The Sky God Tengri is a gentle and just god, one who would never incarnate into such a man, a man who executed thousands, including children. That was not the way of our ancestors, nor is it our way."

Lee thought of the siege of Baghdad, that had taken place in 1258, 60 years after the migration of the Abbot's ancestors to this planet. The Mongol forces had sacked the city and killed the 500,000 inhabitants, most of whom were civilians. She decided not to share that bit of history with the Abbot.

"At first, I thought you were Buddhists," said Lee. "When I saw that you served meat, I knew you were not. But the temple looks and feels like a Buddhist monastery."

"There *was* a Buddhist temple within our community when we came here. We did help them rebuild their house of worship," said the

Abbot. "Over time, they died out. As you know, their monks do not marry, and the few Buddhists who were not monks were just not enough to keep their community going."

He paused and said, "Now, the temple has become a place of peace, study and prayer. Our beliefs and practices have changed over the centuries."

"So, it is a Tengrist temple?" asked Lee.

"No," said the Abbot. "We still acknowledge the Sky God, but we don't worship any deity as such. We believe Tengri is a manifestation of nature, not an actual person. Rather, we acknowledge that we must live in accordance with reason and in harmony with natural law."

"So Daichi Tengri came and took away the whole population?" asked Lee.

"Not the whole population; only hundreds," responded the Abbot. "Many stayed behind. Those who stayed are our kin, and we hope and pray they survived the Khan. When we arrived on this planet, we named it *Dukha* in honor of their memory."

"I was confused, but now I think I understand. I know for a fact that the Dukha are still there. I visited the region as a teenager. At the time of my visit, there were only a few Dukha families remaining, maybe 20. For many centuries, they were believed to be lost to history after the time of the Khan, but they were rediscovered in the 21st century. They are referred to as the *Lost Tribe*."

"That makes sense," said the Abbot. "Only a few hundred remained. Temujin probably ignored them, and they may have hung on, unnoticed for generations. I wish them well."

"One question," said Lee. "Why do your people believe that I am the Goddess Daichi Tengri? You must know that is not true."

"I *do not know* that is untrue," said the Abbot. "Nothing in our belief system would disallow that gods do exist, or that they cannot be reincarnated into human form, even without the knowledge of the host person. To presume certain knowledge on this point would be hubris."

Lee shrugged her shoulders and said, "I really can't argue with that. I can only say for sure that I am no goddess."

"Goddess or not, you may have a role to play in our future." He paused as if deciding to continue, then said, "There is more. It is said that before Daichi Tengri left the people, she vowed she would return."

"OK, that is a common theme in many religions," said Lee. "That a sacred or semi-divine person would return in the future. It doesn't mean it is me."

"No, of course not," said the Abbot. "Not by itself. But there is more."

"Tell me, please," said Lee.

"The monk who met you, Guyuk," said the Abbot. "He had a premonition of your coming 30 years ago. He said he saw a vision that you would crash from the sky into that meadow. He brought his report of that vision to me, and I told him it was a true prophecy of your return."

"What could make you so sure?" asked Lee.

"Turn around, Captain," said the Abbot. "Tell me, what do you see?"

Lee turned around to see a huge tapestry on the wall behind her which she had not noticed when she entered. The tapestry showed a stylized image of woman in bloodstained black armor with large black wings descending from the sky. In the background was what looked clearly like the contrail of a spacecraft burning through the atmosphere. The face of the woman was unmistakable. It was Mei Ling Lee.

"That tapestry," said the Abbot, "is 1,200 years old."

CHAPTER 36

A SACRED PLACE

As Lee walked through the wooded hills, she chose her directions seemingly at random where the path split. But she did so with confidence. Her Marine Corps training had permitted her to understand terrain. Thus, she easily and unconsciously kept track of her location relative to the monastery.

But as time passed, she realized she wasn't taking random turns in the various branching paths. Instead, she felt she knew these woods. This was more than a sense of déjà vu. In déjà vu, one saw a scene and the scene looked familiar, as if previously experienced. Here, she felt she knew what was up ahead of her, and she seemed to be drawn toward whatever was coming.

At one point, she came to a fork in the path and stopped as if confused. Something was wrong and she couldn't tell what it was. She turned around, looking back the way she had come. She saw Sarnai about 100 paces back down the path she had come. Sarnai stopped respectfully, as if waiting to see what Lee would do.

Lee stopped, closed her eyes, took one deep breath, and then slowed her breathing. As she had been taught by her father in martial arts training, she reached out with all her other senses. She heard the wind in the trees, the sounds of small animals rustling in the vegetation, insects buzzing, the feel of the sun, the breeze on her skin.

Then she opened her eyes, and abruptly turned to her left and stepped directly into the heavier brush where there was no visible path. She was trained by the Marine Corps to move through heavy vegetation, and she did so now carefully, but with ease. She could hear Sarnai behind her, moving with less ease as she tried to keep up with Lee.

After about 15 minutes, Lee broke out into a clearing about 100 feet in a rough circular shape. In the center was a statue carved from a rock outcropping. It was clearly ancient, and partially overgrown with vines. Her sense of déjà vu, or whatever she was feeling, was intense. It was more than a feeling of familiarity. It was a remembrance; she had been here before. She *knew* this place.

As she approached the statue, she saw that it was the centerpiece of a shrine, with a small altar in front. The statue was about 18 inches high. Set on the dais, it reached to about chest high on her frame. It was covered in vines, and the face of the statue was obscured. She reached out and began to pry the vines away with difficulty. Because they had been there so long, the tendrils had grown into the stone.

"Are you sure you want to do that," said a voice behind her. Lee turned and saw that Sarnai had followed her into the clearing.

Lee looked at her inquisitively and asked, "Is this a sacred place? Is touching it a desecration?"

Sarnai shook her head and said, "It is not a desecration if *you* touch it."

"I don't understand," said Lee.

"You will. Please continue."

Lee went back to the task of clearing the vines off the face of the statue. As the face began to be visible, Lee's heart began to race, and she hurried in pulling the leaves and vines away. At last, there it was, the unmistakable image of a young woman, Asian features, long jet-black hair, penetrating eyes, dressed in battle armor. It was herself. An inscription read:

When you see this, you will know yourself.

Instantly, Lee was transported to another time, but not another place. She was in the clearing, kneeling before the altar. The statute was clear of debris; it shone brightly in the morning light. Lee saw that the clearing was pushed back several hundred feet, and the grass was neatly trimmed.

Lee reached out and touched the face of the statute and immediately found herself back in the present. The face was once again stained with age and marked by years of vines growing into the rock.

Lee turned to Sarnai and said, "What is this place?"

"It is a shrine to the Goddess Daichi Tengri. In times past, pilgrims came here to pay respects and pray for protection of the Goddess of War."

"Did you know this was here?" Asked Lee.

"Yes, of course," said Sarnai. "Everyone at the monastery knows of it, though no one had been here in years. I haven't been here since I was child."

"Why does no one come here anymore?" asked Lee.

"Daichi Tengri was to return, but that was a millennium ago," Sarnai shrugged and spoke. "Hope fades after so long."

"And what do you believe, Sarnai?" asked Lee.

"I believe you fell out of the sky, exactly as prophesied," said Sarnai. "I saw you find this hidden place on your own, without help or direction. Your arrival is an omen, whether or not you believe you are the Goddess."

CHAPTER 37

THE PATIENT WARRIOR

At dusk, Lee and Sarnai walked the several miles together back toward the monastery. Lee noticed there were places where the brush had been trampled, and evidence of rough campsites were spread about at various places within sight of the monastery. She knew from her training in tracking that these camp sites, and the associated pathways leading to and from them, would mean about 30 to 50 individuals had stopped there. The camp sites were a mess, however. No effort had been made to clean up, much less hide the evidence of a force. She saw discarded animal bones, burned areas where fires had been lit, and piles of stool positioned just outside what must have been the perimeter.

At one point, Lee turned to Sarnai and said, "What is this all about? Surely not the constabulary guard? No military force would leave such a mess."

Sarnai turned away from her, clearly unwilling to answer. Lee shrugged and thought, a mystery to be answered at a future time.

On the way down from the hills, Lee noticed smoke coming from the valley below the monastery.

"What's down there?" asked Lee.

"That's the village of Dahei," said Sarnai.

"Let's take a look," said Lee.

As they approached Dahei, Lee noticed that the town backed up to the cliffs upon which the monastery had been built, providing a natural defense in that direction. She saw that a series of revetments had been erected on the three remaining sides of the village in concentric circles with openings offset from each other to protect the town facing the valley. Along the top of the revetments were walkways on the sides facing the village. Some of the fortifications were old, probably ancient. Areas of the older parts of the wall had been patched up with mortar and mud. Other parts of the wall were recently erected. She immediately recognized the arrangement designed to create a maze through which attackers would need to traverse while being exposed to fire from above, arrows perhaps?

The ramparts were manned by what she assumed were the members of the constabulary guard. The soldiers were wearing body armor, with metal helmets. They were spaced on the wall every three meters, about half of whom were clearly archers, their composite bows slung over their shoulders. Every one of the soldiers carried short swords, either in scabbards at their waists, or strapped across their backs. She saw no firearms and assumed the community had not reached that level of technology.

When they came to the gate, they were challenged by a voice that shouted, "State your name and business for the village."

Sarnai replied in a clear voice, "I am Sarnai, Prioress of the monastery, with a guest of the Abbot."

After they had passed through the gate, Lee saw they had entered a narrow passageway that led them through several twists and turns before it opened into a courtyard of sorts with high walls on the sides with what were clearly firing portals set in the walls several meters about the ground level.

"Why does a small village need defensive fortifications?" asked Lee.

"That is a matter for the commandant to answer," said Sarnai, not meeting Lee's eyes.

"But you know more; tell me. If I'm to be a goddess, you need to do as I say," Lee said with a smile.

"The Abbot did not tell you all," said Sarnai. She stopped walking and faced Lee. "Daichi Tengri didn't just save our ancestors from the Khan. She brought us here with a mission. She said in the far future, there would be a threat. That we should defend this planet. Not just ourselves, but the planet itself."

She paused as if searching for the right words. "We are no longer religious in the way our ancestors were. But we do feel that this place, this planet, is sacred. And we will defend it against any who come here with evil intent."

Lee remained silent as she tried to absorb what Sarai had just said. *Defend the planet?*

Once inside the village, Lee saw that it was functioning at a basic subsistence level. Brick and wood had been used to build mostly single-story structures. The streets were roughly paved, and there was no electricity being used, only the occasional gas lantern showing through darkened windows.

At first, she noticed the eyes inside the building looking out at her. Before she had traveled a block, people were coming out of their houses and businesses to stare openly at her. She heard whispers, and the occasional "Daichi Tengri."

"Is there a tavern where we could get something to eat?" asked Lee.

Sarnai paused, shrugged and said, "If that is what you wish."

"Is there a problem?" asked Lee?

"No," responded Sarnai. "But you will cause quite a stir by your presence."

Lee stopped in mid-stride and said, "I don't have any currency."

"Not to worry, Captain. I have money. But I think you will find they won't let you pay."

Sarnai led the way, and they walked down a narrow alley. Lee wondered why the buildings were so close together. She surmised it must be because the walls of the village were constructed centuries ago, and then as the population grew, the buildings became more crowded.

As they turned a corner, a child came up to them. *She was about eight years old*, Lee thought, dressed in rough but clean clothes. Again, Lee noted the bright blue eyes, something she had rarely seen combined with Asian features. She knew it occurred occasionally when a long-suppressed gene from a European ancestor emerged in a single individual. But she had never seen it in an entire family, much less a whole community.

Lee stopped and looked down at the child. The girl asked in Mongolian, "You are the Warrior Goddess?"

Lee smiled and replied, "I am Mei Ling. What is your name?"

With eyes wide, the child said, "I am Bayalag, my lady."

"It is a good name," said Lee. "It means good fortune."

The child nodded and said in a rush, "Can you fly? Where are your wings?"

Lee smiled and looked at Sarnai who shrugged and addressed the girl. "Young lady, our guest has to be on her way. Can you say goodbye to her?"

The child placed her palms together in front of her face, bowed and said, "May your golden way be filled with prosperity." Then the girl turned and fled down the narrow alleyway.

Lee looked at Sarnai and said with a smile. "I'm afraid the villagers are going to be disappointed with me. I don't even have any wings."

Sarnai shrugged in an odd way and was silent.

"What?" Lee asked.

"You *do* have wings," replied Sarnai in a steady voice. "Your battle armor has wings, and you rode them from the heavens."

About fifty meters further on, they came to an unobtrusive building, much like the others in the village. Its only distinction was a small hand-painted sign over the door that read, *Three Taverns*.

Lee asked, "Are there three taverns in this town?"

"No," replied Sarnai. "The name is centuries old. No one remembers why it's called Three Taverns."

They entered and Lee was immediately hit with the smell of cooking food. She could hear conversations. As she looked around, she saw dimly lit circular tables with patrons sitting in groups of three or four. At the far end was a bar and behind that was a tapestry showing wild animals, hunters and castles.

A moment after they entered, the room fell silent.

Sarnai led them to a small table near a window and they sat down. A man with an apron came to the table, bowed first to Sarnai and said, "Welcome, Madam Prioress." And then he turned to Lee, bowed deeply and said, "Blessing and mercy upon a warrior."

Sarai said to Lee, "Captain Lee, this is Master Khulan, proprietor of this restaurant."

Lee said, "Master Khulan, thank you for the sanctuary of your house."

Sarnai said to Khulan, "We'll have Buuz with tea, please."

"Right away, ma'am," he replied and walked back toward the kitchen.

Lee noticed that the conversation in the restaurant had started again, albeit with a lower volume than when they had entered.

The food came quickly. Buuz turned out to be dumplings stuffed with a sweet-tasting meat. The tea was strong and, Lee thought, *delicious*.

Tell me about the fortifications around the village," said Lee.

"Again, that is a question better put to the commander of the guard," replied Sarnai.

Lee said with a steady gaze, "I will, but I'm asking you what you know. You've lived here all your life, some of the construction is ancient, some of it is new."

Sarnai shrugged and said, "One of the things one learns from growing up in the monastery is to speak only on subjects one is versed in."

"You said that you thought I was a goddess. *Answer my damn question*," Lee said with her cold face.

Sarnai was silent for a long moment, and Lee could see she was struggling with something as she blushed slightly.

"Very well," she said at last. "The prophecy of... the return of the Daichi Tengri did not only predict that she would return."

"And..." said Lee with some impatience in her voice.

Sarnai was obviously struggling to decide what to say or not to say. "The prophecy also states that she will return at the time of ultimate need." Then she looked away as if she regretted making that statement.

Lee thought a moment, and then said, "So there is a threat to the village, and I assume also to the monastery. Tell me what you know about..."

Just then there was a shout from the street outside the tavern.

CHAPTER 38

THE NEPHILIM

Lee could hear the growing shouts of people, and the sounds of running outside the tavern. A bell in the near distance began ringing. The men in the tavern simply got up and began walking toward the door, purposefully, but unhurriedly. Before the first of them got there, the door burst open, and three men entered in full battle armor. The occupants stopped in their tracks and stepped aside for the trio as they came directly to Lee and Sarai's table.

In the lead was Noker Aruban, the commander of the guard. He looked directly at Lee and said, "Captain Lee, you are needed."

Lee cocked her head and said, "Yes? Tell me."

"The village is under siege. We need your help. We've brought your battle armor and weapons from the monastery." Aruban nodded to the soldier on his right who unslung a canvas sack and held it out to Lee.

Lee took the offered sack, opened it and saw that indeed her armor and weapons were inside, all neatly stowed. As she drew it out, she noted that someone had cleared off the blood, dirt and leaves; and made a good effort at polishing the composite metal. She smiled inwardly, thinking that

Marines typically spent a good deal of time intentionally scuffing up the exposed metal on their armor so that it would not be so visible to any enemy with a sniper riffle.

Quickly, she donned the armor and performed a quick function check on her weapons.

"OK," she said looking at Aruban. "Show me the enemy."

The five of them moved into the street. It was a scene of barely surpassed chaos. Civilians were moving quickly, carrying bags in their hands or on their backs, all moving toward the direction of the monastery. Soldiers in battle gear were moving toward the walls of the village, some individually, and others in squad-sized units of about a dozen each.

Lee followed Aruban who set off at a surprisingly quick pace. She noted Sarnai stayed with her, while the two other soldiers brought up the rear.

She whispered to Sarnai, "Are you coming for this? It looks like combat."

Sarnai said only, "I am the Prioress; my duty is at the wall. Also, I am to stay with you."

As they approached the wall, Lee could hear the sounds of shouted commands, clashed metal weapons, and the swoosh of arrows being fired in volleys. Above the wall she saw the flickering light of fires from the opposite side.

Aruban took them up a stone stairway cut from the inner blocks of the defensive wall. As they crested the top, Lee looked across the three concentric walls and saw the outermost wall was a teaming mass of bodies, moving, and writhing. Beyond the outer wall she could see multiple fires and what sounded like chanting.

She turned to Aruban and said, "What is the tactical situation?"

"The enemy is armed with arrows, spears, short swords, and knives. They are also capable of fighting unarmed and will use their teeth and nails if needed."

Aruban paused, looked at Lee and then turned his gaze to the battle on the outer wall. As she waited for him to continue, she watched the battle more closely. Clearly, the outer wall had been scaled and the defenders overwhelmed.

She couldn't see the attackers clearly, but she heard them. They let out guttural cries that did not sound entirely human. Just then one of the attackers on the outer wall rose to his full height, a giant figure, well over three meters. He held a struggling Mongol defender by the hair, several feet off the ground, and roared with a mixture of triumph and ecstasy. With the blade in his other hand, he severed the man's head from his body with a quick movement, letting the body fall away while holding the head triumphantly in his hand. And then, to Lee's horror, he brought the face of the decapitated man to his own and took a bite with enormous teeth.

She looked back at Aruban unable for a moment to form words, then she said. "Not human."

"I think it depends on what means human," he said.

"What are they?" she asked.

"They are Nephilim, fallen ones," he replied.

Lee looked at him quizzically. "You mean like from the Jewish bible? Offspring of angels and humans?"

"I don't know that book," replied Aruban. "When we first came here to this planet, our ancestors encountered the giants. Someone, perhaps a Christian or a Jew among us, named them."

"Their strength?" asked Lee.

"Not clear," responded Aruban. "We've never seen them in anything like these numbers. And they have never made a full onslaught against the walls before."

He paused as if thinking and then said, "Before they were overwhelmed, the commander of the watch sent a message saying there were ten fires outside the walls beyond the range of our arrows. In normal circumstances, that would mean about ten groups of 80 to 100. It's a guess only, but I'd say we are looking at a maximum of 1,000."

145

Lee said, "When we were out walking the hills today, I saw signs of old campsites. Was that from the enemy?"

"Yes," he replied. "They sometimes go into the hills, and we see an occasional raid on the reindeer herds. They have attacked farmers and some of the outlying settlements. More so in recent times. Thus, the new construction on the wall."

"Here's what I don't understand," said Lee. "I was out there only a few hours ago. I didn't see any sign of recent activity, certainly not 1,000 of that size. No recent tracks, stool, or crushed brush. How did I miss them? It doesn't make sense. I'm trained in reconnaissance; it was my primary job until two days ago."

"Tunnels," he said simply. "They can dig them quickly. When we find these tunnels, we destroy them and fill them in. But it doesn't seem to deter them much, if at all."

"Are they all so large?" asked Lee.

"No," responded Aruban. "There seems to be several types, some are smaller, maybe five feet tall. They are very fast and capable. Do not underestimate them. Then there seems to be a digger class. They move on four feet, maybe seventy pounds, and resemble a blind digging animal. We don't know whether these are domesticated animals or just a variation of the species. They can attack if pressed, but that doesn't seem to be part of their normal tasks. Basically, they are as dangerous as any wild animal with teeth and claws. Mostly, though, they dig the tunnels."

"And your strength?" asked Lee.

"We had 158 including the auxiliaries from the village," replied Aruban. "The entire first watch has fallen, so now 130."

"What is your plan of defense?" asked Lee.

"In order to get to each successive wall, they will need to descend to the space between the walls. We plan to fire into the space between the walls as they advance," said Aruban. "We have archers with thousands of arrows. We also have an incendiary gel we can pour onto the defenders as they climb the wall."

Lee shook her head and the commander asked, "What?"

"That will only work if they need to enter the space between the walls," said Lee.

"And they must," replied Aruban. "Unlike you, they cannot fly across that space."

"Not fly," said Lee. "Look."

Aruban looked up and saw what she meant. The Nephilim, now in complete control of the outermost wall, were extending poles across the expanse between the outer wall and the next wall.

He swore as they watched one of the attackers, a smaller one, jump onto the pole and begin to scamper across it at alarming speed. One of the defenders fired an arrow piercing the attacker through the torso. To Lee's amazement, the thing kept climbing with the arrow clearly protruding from its chest. The next volley of several arrows finally did the trick, and the attacker fell into the space between the walls. But already several other poles had been extended, and Lee could see multiple bodies scampering across the expanse.

Aruban swore again and turned to shout orders to send reinforcements to the second wall. He turned back to Lee, a look of dawning panic on his face.

"Can you help?" he asked.

Lee scanned the entirety of the battle space for several seconds. Then she turned back to the commander and said calmly, "You have less than 100 attackers on the wall. Maybe, you can hold this group off. But if they have 900 more in waiting, this defense will fail."

"What do you suggest?" he asked.

"Why, attack of course," she replied.

And Aruban, looking at her in the flickering light of the fires, thought he saw a wicked smile.

CHAPTER 39

A NIGHTTIME PATROL

Lee, Aruban, Sarnai and a squad of ten soldiers assembled by an ancient door nestled under an awning of the wall on the West side nearest the cliff base.

"This is it," said Aruban to Lee. "These are the best."

Lee had asked for a squad of no more than 12 soldiers, preferably trained in reconnaissance, raids and ambush. Aruban had called for the scout platoon leader, Lieutenant Andras, who had selected nine other warriors best suited for the mission.

Lee looked them over in the dim light. She didn't like going into combat with soldiers she didn't know and had not trained with. But these would have to do. And they looked like solid, tough-looking men, who carried themselves with the gravitas of experienced warriors. Calm, alert, ready but not nervous. They were lightly armed with bows and arrows, a short sword, and knives. They wore light armor, and their hands and faces had been blackened, so that there was no reflective surface.

Lee turned to Sarnai, who was now armed with bow and saber and wearing a light leather armor vest, and said, "I would prefer that you not come. Your presence isn't necessary for the mission and you may just get in the way."

Sarnai smiled and spoke. "I must accompany you; those are my orders from the Abbot. And I have been trained by the monks sufficiently that I will not present a burden."

Sarnai turned to commandant Aruban. "Alexi, do you have an objection?"

Aruban said, "No, Prioress. You are of course a welcome asset on this or any mission. None from the monastery has ever shrunk from battle. I trust you will look to the captain's safety as a priority."

Sarnai looked at Lee while she responded to the Commandant. "I suspect Captain Lee can manage her own affairs. My priority is the mission. I will of course come to the aid of any in danger should their need be great."

Lee was seeing a side of Sarnai she hadn't noticed before. She seemed taller, and moved and spoke with a commanding, confident air. Clearly the commandant, and it seemed the other soldiers, deferred to her is some way. It was more than social status belonging to the Prioress. They seemed to genuinely respect her.

Lee said to the group, "Gather round." As they formed a semi-circle with her facing them in the center, she quickly laid out the plan.

"I am Mei Ling Lee, Captain in the Alliance Marine Corps," began Lee. "I am the commanding officer of this mission. Second in command is Commandant Aruban and third is Lieutenant Andras."

She looked around at each of the faces, to make sure they understood who was in command and what was the line of succession if they started to take casualties. She knew from training and experience, better to make sure everyone on an operation knew who was in charge and what they intended to do.

"The mission of this operation is to conduct a reconnaissance to determine the location and composition of the enemy's command and control, to attack, impair or destroy that command element, and to return safely within the walls."

Lee held a two-foot stake to draw a rough depiction of the areas in the dirt at her feet.

"Execution: we will exit the walls of the town, *here*," she indicated a point on the map. We will initially travel in file formation. Lieutenant Andras, you will designate a scout to take point to navigate the route." She looked at Andras and he nodded. "I will be immediately behind the point, with Commandant Aruban. Following will be first squad, then Lieutenant Andras with the special weapons team, then second squad."

Lee stopped and looked around, both to emphasize the instructions and to ensure they had understood. She saw all eyes on her, with no sense of confusion or resistance. These were professionals all right.

"We'll move along here," she indicated on the map with her stick. "When we get to the mouth of the stream, we will set up an objective rally point, that's the place we return to if we become separated. From there we should be able to see into the enemy's rear. We'll take stock and decide what action to take next."

Lee looked up at the faces in the group to see if there was any sign of doubt or question. Not finding any, she turned to Aruban and asked, "Do your scouts know how to clear danger areas?"

"Yes, Captain," he responded. "We send two soldiers into a danger zone: an open area, or road. They cross the zone and quickly scout the area beyond. Once safe, they will signal for the team to follow."

Lee nodded. She was gratified these scouts used a procedure similar to that used by the Marines. She supposed it was a universal precaution when entering an unknown area. The tactical measure was designed to ensure no ambush awaited the entire squad. Ideally, the two-person clearing team would locate the ambush and the squad could either avoid it all together or attack the enemy on more favorable terms. Less than ideally, the two advance guards would trigger the ambush, and at least then the whole patrol would not be lost.

"Any questions?" she asked.

One of the scouts raised a hand. Lee nodded to him and said, "Yes?"

"You have weapons with you that are far beyond what we or the enemy have." It was a statement. "Will you use them on this mission?"

Lee looked at the man, then at the rest of the patrol, then finally at Aruban. "By my law, I may only use these weapons in self-defense or to protect Alliance personnel. So, I think the answer to your question is no, probably not. I am sorry for that, but it is the rule by which I am bound."

The man nodded in satisfaction. He had asked a direct question and received an answer.

When there were no other questions, she said, "Good, Let's go."

CHAPTER 40

OBJECTIVE RALLY POINT

Aruban gave a brief command, and the patrol spread out into a tactical formation. Each soldier took a knee.

Then Lieutenant Andras sent two men through the opening in the wall. A minute later she heard a voice, "All clear."

Quickly, one by one, the soldiers moved through the opening in the wall. When it was Lee's turn, she was surprised to see that it wasn't just an opening in the wall, but rather a tunnel underneath it. Toward the end, she had to crawl on her hands and knees. Involuntarily, she felt uncomfortable being in such a dark and confined space. The tunnel looked ancient, and she doubted anyone had thought to ensure that the roof of it had been reinforced or even checked in centuries. Alas, there was nothing to do but push on.

Soon enough she was out in the open and, despite the danger of the pending mission, she felt relief breathing the open night air. The night sky was spectacular with stars. This planet was closer to the galactic center and thus the stars were much more prominent than on earth. Off to her left, she could see the bonfires of the enemy. She could hear them as well, guttural shouts. There was no sign that the Nephilim had set up any sort of guard or perimeter. Clearly, they hadn't thought of themselves as vulnerable.

The patrol moved quickly along the tree line, and Lee was gratified that the soldiers moved quietly and didn't bunch up as they would have had they been inexperienced or untrained.

Behind her, she could hear the fighting along the wall. By now the attackers would have gained control of the second of the three defensive walls. The fate of the village now depended upon the success of their mission.

In about 20 minutes, the patrol arrived at the place Lee had selected for the objective rally point. It was a small clearing, adjacent to the open area that allowed some concealment, but also good observation of the enemy's positions. Lee was gratified that the scouts knew how to occupy a rally point. The two soldiers in the lead entered the clearing, quickly determining that it held no hazards. One remained on the far side while the other returned to the rest of the patrol and signaled that the position was clear. The remainder of the patrol entered quickly but alertly, taking up positions on the roughly circular perimeter, in prone positions facing outward.

Lee signaled to the commandant and the lieutenant to come to her in the center. When they arrived, she said, "Commandant Aruban, please select two of your best scouts. The four of us will conduct a further recon of the enemy positions. Once we have the information we need, we will return here and take the remainder of the force forward."

Both Aruban and Andras nodded their understanding. Andras then turned and whispered the names of two of his scouts who quickly joined them in the center of the circle.

"Commandant," said Lee. "Am I correct that you have a method of summoning the patrol forward in case we need them immediately and there isn't time to return for them?"

"Yes, Captain, I have a horn," he said pointing to a curved device on his belt. "If blow it, they will come."

"Won't that alert the enemy?"

"Not necessarily," said Aruban. "It makes many sounds. The one I will use mimics the sound of a reindeer in heat. My people will know it; others likely not."

CHAPTER 41

INTO THE ENEMY'S REAR

Lee led the small group further to the rear of the enemy's position. It took about 30 minutes to get where she wanted, about 100 meters behind what she considered the center of their line. There were actually only five bonfires still burning. She assumed the other five that had been identified by the watch captain earlier had been left to burn themselves out when the bulk of the enemy had attacked the wall.

This was good news of a sort. Lee estimated that only about 100 of the enemy had remained behind, while about that same number had attacked the wall. Earlier estimates had the number remaining behind at about 900.

Lee looked closely at the center bonfire. There seemed to be some kind of structures immediately in front of the fire that the other bonfires didn't have.

Lee turned to Aruban and asked, "Can you make out what those structures are?"

"Those are punishment racks," he said. "The Nephilim use them to torture captives, or their own members that have been judged as cowards or criminals."

"Let's' get a closer look," she said.

As they got closer, they moved ever more cautiously because the light of the fires made it more likely that they would be detected.

Finally, when they got to about 100 feet of the structures, Lee had to suppress a gasp at what she saw. Each rack held a figure suspended in what amounted to a crucifixion position about four feet off the ground. She saw with horror the four missing members of her crew Zakany, Danner, Ali and Odessa.

Lee turned to Aruban and said simply, "Those are the members of my crew. I need to get them back."

"Yes, I can see," he said while nodding. He took a moment in contemplation and Lee let him think without interrupting.

He said, "The best solution is to conduct a double tactical envelopment, attack from both sides of the formation at the same time, while sending a small crew to extract the four prisoners from their confinement."

Then he shook his head and continued, "But we don't have enough troops with us to make that a success, and we don't have time to get more. The four on the rack could be killed at any time, and some or all of them may already be dead."

"Any options?" she asked.

Aruban looked steadily at her and said simply, "Your weapons."

CHAPTER 42

RESCUE OR RECOVERY

The plan was simple, partly because it was the only plan that could work in the time available. A more complicated plan would require more people, more resources and more coordination. None of those elements were available.

Lee did not like the plan, even though it was of her own making. She would stay well back, laying prone on a slight rise in the terrain. She would be sniper and provide over-watch. It was the only way. Sarnai had decided to go with her, and would serve to protect Lee while she was shooting, should that become necessary.

The Marines favored percussion projectile weapons because they could be used in almost any environment, including in a vacuum. The weapons were reliable, they could be fully functional when wet, dirty, hot or cold.

Lee had made the decision to use her firearms. The Alliance had a hard-and-fast rule that prohibited using advanced weapons on planets that did not have such weapons. The exception was that firearms could be used if necessary to save the lives of Alliance personnel. Upon reflection, Lee had decided that the current situation met that sole exception. Her four team members were all part of an official Alliance military mission. Clearly, their lives were in immediate peril.

Lee knew that if there was ever a review of her actions on this day, someone would likely raise the objection that she had used the prohibited weapon as part of a coordinated offensive scheme, and had furthermore fired first without an immediate provocation.

So be it, thought Lee. She had been charged and court martialed once before for similar violations. On that occasion, she had been acquitted of all charges, and later even been awarded a medal for her actions. But that was a different time, and she had had powerful advocates on her side. Maybe next time she wouldn't be so fortunate.

Again, whatever might come, she would do what was necessary to save her friends. And there was every possibility that this would end badly for all of them, and thus no need for a court martial.

She had two weapons: a carbine and a handgun. Both weapons fired projectiles, much like her ancestors on earth had used centuries ago.

Aruban would lead the extraction team. It would be small, only four men. The remainder of the force would make a flank attack under the command of Andras. The goal was to create a disturbance and draw the enemy to the flanking attack. In the confusion that was expected, Aruban, supported by Lee's overwatch, would take back the hostages.

The hard part, of course, was the retreat back to the relative safety of the village walls. Once the attack had been launched, all surprise would be lost. It was unlikely the four captives would be able to move under their own power. That meant carrying them, which would definitely slow down any retreat.

Lee looked through the night scope mounted to her carbine. She was tempted to survey her friends for signs of life, but she knew the mission required her to focus on the targets. The hostages were either dead or alive. It didn't matter for the conduct of her part of the mission. If they were alive, they would do everything to save them. If they had already succumbed, she would extract their remains for burial.

Through her scope, Lee began to get a better picture of the scene. Luckily, the four captives were all in a single place, each tied to a configuration of wooden beams. That would make the extraction easier than it would have been had the four been spread out among the bonfires.

There was no obvious guard force. She knew from Aruban that the Nephilim came in three sizes. In her own mind, she had named them *Big Brother*, for the three-meter-tall giants, *Little Brother* for the smaller five-foot kind, and *digger* for the mole-like creatures.

The enemy seemed to wander about, sometimes approaching the captives, but most often just passing by. None of the captives showed any sign of life. Lee hoped they were unconscious rather than dead.

She used the laser rangefinder in the scope to check the distances. The captives were about 150 meters away. The various enemy in the area ranged from 120 to 200 meters.

Then Lee put down the weapon and checked her ammo. Three hundred rounds. She hoped it would be enough. She would have to be careful, precise, making sure every round counted and nothing was wasted. She thanked the Marine Corps for sending her to sniper school. She was nowhere near as good a marksman as some of her marine snipers in the 75th Recon company. But she was competent.

She checked her sidearm. It was functional at short range only, designed for close combat in situations where the longer carbine couldn't be used. For it, she had 50 rounds.

Finally, she drew her knife from its scabbard at her waist. It was a Marine Corps issued Ka-Bar. Weighing just under a pound, it was a foot long with a seven-inch blade. It had a comfortable grip made of a composite material, and a short hilt designed to protect the hand when thrusting against resistance. In training, she was taught that the Marines had used this same knife all the way back to the wars of the 20th century. Lee had used it twice in the six months she commanded the 75th; both times during night raids into terrorist camps where she used it to silence and kill guards. She found it a good weapon for close fighting for both thrusting and slashing.

As an optional feature, the knife could be used affixed to the barrel of the carbine as a bayonet. Lee had practiced with the bayonet but had never had the occasion to use it.

She recalled that once during her time in the Navy, a young provost guardsman who was protecting her command center from attack by masked raiders, had used the bayonet as a last resort when all his team had been killed and he had expended the last of his own ammo. He had fixed his bayonet and launched a one-man, furious counterattack. He died from multiple gunshot wounds, but not before stabbing a final enemy in the face.

Admiral Chambers had posthumously awarded the young man the Navy Cross for heroism. In the aftermath of the battle, Lee had come across Chambers while he was reviewing the video of the guardsman's final moments. It was the only time she had ever seen the old warrior cry.

Just then she heard the unmistakable sounds of the Mongols attack on the enemy's flank.

CHAPTER 43

EXTRACTION

Lee was surprised at all the commotion the small Mongol element made when beginning the attack. The sounds were a combination of yells and what she assumed were horns sounding an attack signal. In addition, they had lit several fires spread out over several hundred meters. In all, the combination was designed to make it look like a larger force was attacking.

The effect on the enemy was immediate. Every single one of the Nephilim that she could see stopped and turned toward the sounds of battle. One of the giants near the four captives opened its mouth and roared. It was that same sound Lee had heard on the wall a few hours before. All across the enemy's position, the howls rose up.

Through her night vision sight, Lee saw that the Nephilim began moving toward the Mongol flank attack. She just hoped it was enough. She could see Aruban and the other three Mongols moving quickly in a crouched position towards the captives.

Two Nephilim were in her sights. A little brother was apparently speaking to a big brother and gestured toward the four captives. The big brother turned toward the captives and drew his sword. The silenced 30-

millimeter bullet left the barrel of her carbine at 3,000 meters per second and found its target, big brother's head, in two tenths of a second. Lee saw the bullet enter, and watched the creature's head snap over. The shot was spot on.

But then something unexpected happened. The Nephilim staggered, went down on one knee, and *got back up!* The thing howled in what sounded like outrage and pain. Several of the big brothers turned at the sound and began moving back toward their injured colleague.

"*Shit*," said Lee under her breath. This was going to be harder than she thought.

Lee switched the carbine to three-round bursts and fired again into the giant's chest. The impact of the bullets slammed into him and drove him back. But still, he didn't go down. Lee placed the crosshairs on the giant's throat and squeezed off another burst. This time there was an effect. The giant's neck exploded, and the head lolled backwards, almost severed. Finally, it collapsed.

Lee knew she was in trouble. It had taken seven rounds to take down just the one Nephilim. Beyond that, with her continued firing she had almost certainly given away her position. She widened the angle on her scope and saw that other Nephilim were being drawn to their downed colleague, and they were examining his corpse and searching around for the source of the attack. The Nephilim may not have been exposed to firearms, but they were familiar with projectile weapons; namely, arrows and spears.

The worst part is that she had fouled up the plan. Rather than moving toward the flank attack, at least some of the Nephilim were now clustering around Lee's first target. And that target was only a few yards from the captives. For the plan to work, Aruban and his three soldiers needed the area around the captives to be clear of Nephilim. No way could they fight through half a dozen of the monsters.

Lee switched to a combination of bullets with varying characteristics. Her three-round burst would comprise an armor piercing round, followed by a hollow point, then an explosive round. The idea was that the first round would penetrate the heavily muscled exterior, the second one would enter and then expand the wound, while the final round would enter the enlarged wound and then explode inside the target.

The weapons she was using were adapted for both sound and flash suppression. This was a design element that made it difficult for an adversary to see the muzzle flash or hear the weapon firing. But it wasn't perfect. Some light and some sound always escaped. Lee was firing from a relatively exposed position because there was little vegetation to conceal her firing.

She decided to fire one more time and then move to an alternate firing position. She took aim at one of giants. She waited for him to turn his face toward her, and she squeezed off another three-round burst. This time the effect was catastrophic. The giant's face exploded, and great pieces of its skull were blown up and backward. It collapsed. Lee said to herself: *that one won't be getting up.*

As soon as she saw the impact, she was moving. She called to Sarnai, "Let's go. Gotta find a new spot to shoot from." Sarnai made no reply but simply moved quickly and quietly with Lee. *Impressive,* thought Lee. Sarnai seems to at least keep her cool under combat conditions.

Lee picked a spot about 100 meters to her left, and she laid down prone to take up a firing position. She could hear the battle raging off to her left, and she hoped Andras and his men were holding on.

She picked up the scope again and saw that the Nephilim's movements were getting more coherent and coordinated. A group of five of the giants had formed a battle line of sorts and were moving directly toward her previous position. Clearly, they had seen or heard the firing. She suspected their senses were enhanced for night fighting.

They had left only one guard with the captives and Lee realized that their plan might work after all. The Nephilim apparently hadn't seen Aruban and his men, but rather were intent on reaching her former position, which was now empty. If she could keep them occupied, Aruban would have a chance to get the hostages out.

Lee waited until the five Nephilim had reached her now-abandoned sniper's nest. She could see them searching the area. Then they did something unexpected. Two of the giants got down on all fours and began bobbing their heads up and down, back and forth. "They're sniffing our scent," she said out loud to Sarnai.

"Yes," said Sarnai. "They are known for it."

"I'll start to engage them now," said Lee. "You should be prepared to move again quickly. Once I start firing, they will probably see us directly."

"I am ready to move with you any time, Captain," Sarnai said. "But we have other problems."

"What?" said Lee.

"Look behind you," said Sarnai.

CHAPTER 44

IMPROVISE AND RUN

Lee turned and saw movement in the direction Sarnai had indicated. She flipped on the night vision sights and saw what looked like a pack of giant moles moving toward their position. And they were *running*.

"I thought the commandant said these diggers would only attack if provoked."

Sarnai shrugged and said, "Live and learn."

Lee realized she was out of time. Both enemy elements would be on top of her position in a few moments, and she could not deal with both at the same time.

"Captain," said Sarnai calmly. "May I suggest you give me your smaller firearm. I might make good use of it under the circumstances."

Lee realized she had to decide right then. Giving a firearm to anyone outside the Alliance was prohibited. But just now she didn't have a choice. If she didn't give Sarnai the weapon, they would almost certainly die or be taken captive. And the mission to rescue her crew would fail.

Lee quickly unholstered her gun and held it out to Sarnai. She said, "The safety is off, just aim and pull the trigger. The weapon has laser sights, here. That will show you where the bullet will go. Aim at that point. Hold it like this," Lee demonstrated the two-hand firing position.

"Got it," said Sarnai with supernatural calmness. She snatched up the weapon and to Lee's surprise, took off and ran on a course perpendicular to the advancing super moles. At first, Lee thought Sarnai was running away, but then she realized Sarnai was trying to draw the creatures away from Lee so that Lee would be free to deal with the giants.

Although Lee hated to turn her back on Sarnai and the moles, she needed to deal with the giants who were now only a few meters away and closing directly upon her position. They definitely knew where she was.

Lee raised her carbine to the shoulder firing position and fired a three-round burst into the leg of the forward-most giant. The impact literally blew the leg off at the knee and the giant slammed face first into the rough ground, giving off an agonized howl of pain and frustration.

As Lee shifted to the next target, she heard the sharp reports of her sidearm behind her. Clearly Sarnai was engaging the moles. Her shots were spaced and measured. Lee was impressed. A first-time user of a firearm often panicked and just fired continuously, wasting precious ammo. Lee heard what sounded like angry squeals, and she assumed Sarnai was inflicting damage.

Lee blasted the next two and they both went down. Three down, two to go. Unfortunately, the remaining two had burst into a run and were trying to close the distance to Lee before she could get them both.

Lee realized she wasn't going to get both Nephilim before at least one of them would be upon her. She leapt up and began moving laterally while firing. She aimed for center of mass now, not trusting a more precise shot to the knee as she was running. To her horror, the three-round blasts to the torso had only a minimal effect on the creatures. She could see them stagger and howl, but they kept coming.

The closest one was upon her; its giant jaws open, showing gleaming fangs. It was close enough for her to smell its breath, a foul stench of rotting flesh. She only had time to turn to face it. As she did, she tripped and fell backward. She landed on her back just as the massive jaws began to clamp down on her. Instinctively, she turned the carbine toward the creature and, without actually meaning to, slid the barrel directly into the creature's mouth.

Lee pulled the trigger and the weapon fired its three-round burst directly into the open mouth. The Nephilim's head exploded rearward and the upper and lower jaws actually separated, the top half spinning away.

Lee tried to roll away, but the barrel of the weapon was somehow lodged in the creature's throat. She had no choice; she abandoned the carbine, rolled out from under the carcass, and came into a crouched fighting stance with her Ka-Bar drawn.

The remaining giant was only about 20 feet away and charging, but Lee didn't wait for it to reach her. She charged the beast, letting out a roar of her own. Her aggressive movement seem to flummox the Nephilim which stopped in its tracks.

Lee feigned a frontal thrust and then dodged to her right, slashing the calf of the giant as she shot by it. She felt the knife penetrate and heard the giant scream in pain. The creature reflexively reached down to grab its injured limb, and when it did, Lee plunged the Ka-Bar into the rear thigh muscle up to the hilt. She yanked the knife free, and stabbed again, this time in the other leg. The creature went down on both knees and let out a hideous wail of what Lee assumed was pain and frustration.

The giant, now on all fours, scampered around to try to face her, snapping its jaws, making a loud, cracking sound. Lee moved quickly away from the jaws and sprung up onto the giant's back. The Nephilim, realizing its peril, frantically tried to reach behind its back, howling in rage and pain. Lee scrambled to a position between its massive shoulder blades, put one hand in the thick hair of its head, pulled back and then reached around intending to cut deeply into the exposed neck. A moment before she pulled the knife across the giant's throat, the beast froze, and let out what to Lee sounded like a mournful cry.

Lee made the cut; she could feel the flesh, muscle and cartilage yield to the wickedly sharp Marine Corps blade. A great fountain of blood spurted out from the wound. The giant gave what sounded like a trumpeted sigh, and then dropped prone and unmoving.

CHAPTER 45

OLGOI-KHORHOI, THE DEATH WORMS

Lee frantically looked around for her carbine and saw a slight reflection of firelight in the dirt a few meters away. She hopped off the now-dead Nephilim, scooped up the carbine and began running in the direction she had last seen Sarnai.

Lee scanned the area with her night scope. She didn't see anything at first. Then she saw it: movement about 100 feet away. She set out at a run, hoping beyond hope that Sarnai was still alive. As the movement became clear, Lee was astonished. Sarnai was calmly walking along what looked like a row of dead bodies. *Had she killed them all, then lined them up?* No. Not at all. The giant moles were all still alive and they were patiently sitting in a row, with Sarnai walking up and down the line, speaking softly to them, reaching out and petting them.

"What happened here?" asked Lee.

Sarnai turned to her, smiled and said, "Say hello to my friends, the Olgoi-Khorhoi."

"I don't understand," said Lee.

"The Olgoi-Khorhoi, the death worms," she said. "They are an intelligent species held captive by the Nephilim. They were not running to attack us, but to escape their own captivity. They recognized me as a member of the monastery, and they ran to me to gain sanctuary."

"Why the shots then?" asked Lee.

"Those were for the Nephilim who were chasing our friends here. I used your gun and later my own knife when they got close enough. Our friends here fought with me, some paying the ultimate price. She gestured behind her, where Lee saw for the first time the several bodies, two of them large, obviously Nephilim, and three other smaller ones that Sarnai and the Olgoi had killed while resisting recapture. Set apart from the bodies of the enemy, Lee saw two smaller bundles of fur.

Lee looked at the line of patient Olgoi, and they looked back at her with sad, intelligent eyes.

"What now?" asked Lee. "Where will they go?"

"They will come with me to the monastery, where they will receive temporary sanctuary while permanent accommodations can be made," Sarnai said. "For now, they will fight with us and help us to return safely."

"Can they speak?" asked Lee.

"See for yourself," said Sarnai. "This is Arslan. She is their leader."

Lee approached the line of Olgoi. One of them, Arslan she supposed, stood up on its hind legs and came forward to Lee. The walk was not awkward at all. Clearly this species was as comfortable on two legs as on four. As Arslan approached, Lee noted intelligent, thoughtful eyes. Arslan looked directly at Lee, reached out, took her hand and sniffed it.

What happened next astonished Lee. Arslan looked up sharply at Lee, dropped her hand, backed up a few steps, snorted and then lowered herself onto her elbow joints, so as to be almost prostrate. Lee looked at Sarnai who shrugged and said, "I have no idea."

Then the Arslan said in a hoarse but perfectly clear voice:

"Daichi Tengri."

All the Olgoi faced Lee and similarly prostrated themselves and repeated in unison, *"Daichi Tengri."*

CHAPTER 46

SLAVE REVOLT

A few minutes later, Lee was running toward the place where the captives were being held. With her were Sarnai and five Olgoi spread out in a surprisingly well-disciplined military V-shaped formation. She could see that Aruban and his three scouts were in a fierce battle against a half dozen Nephilim. They fought with great skill or else they would have succumbed by now. But they were losing, and it was only a matter of time.

Lee wanted to use the carbine to assist but the battle raged around the captives and Lee could not be sure she wouldn't strike one or more if she started firing.

Next to her was Arslan, who suddenly gave out a very loud call that sounded something like the howl of a wolf combined with the trumpet of an elephant. Immediately, her call was answered with similar calls all across the line of battle.

Lee was going to ask what that was about, but it soon became clear. The Olgoi were rising up against their masters, the Nephilim.

"What did you do, Arslan?" Lee asked.

"I told them Daichi Tengri has returned. We have waited 1,000 years," she said in her gruff voice.

Lee didn't respond. She wasn't about to argue that she wasn't Daichi Tengri. Maybe later they would have that conversation. But for now, the misidentification seemed to be working as an advantage to her side.

She saw that the Olgoi were attacking the Nephilim around the four captives. The ferocity of the attack was breathtaking. Dozens of Olgoi attacked every Nephilim. They were extremely aggressive, jumping up onto the Nephilim, ripping and tearing with teeth and claws. The Nephilim fought back, swung huge clubs, or ripped with their own teeth. But for the Nephilim, it was a losing battle.

Lee could see that a group of the Olgoi, about a dozen, had formed a protective circle around the four captives. Toward the end, several of the Nephilim tried to run away, but they didn't get far. The Olgoi relentlessly tracked them and pulled them down under an avalanche of teeth and claws.

By the time Lee, Sarnai and the five Olgoi reached the site, the battle was over. All of the Nephilim had been killed. The Olgoi had formed a broad perimeter now around the sight, facing outward in a protective posture.

Aruban and his team were cutting down the four captive members of her crew, and performing first aid. Sarnai took charge of the recovery effort. She pulled some ointments from her pack and began cutting away at the already ragged remains of their flight suits in order to check for and treat the wounds.

Lee rushed over, scanning quickly to see their status.

"How are they?" asked Lee.

"Three are unconscious and seriously dehydrated," said Sarnai. "They had many puncture wounds, which went untreated."

Lee knew this meant that they had been tortured.

"A woman is conscious, also very dehydrated."

Lee went to Zakany. "Laura, I'm so glad you are alive. God, what you must have been through."

Laura looked at Lee with a weak smile, and said softly, "Roles reversed. Last time you were hurt, and I was stuck on the command ship."

Then Zakany said, "The others. Have they survived?"

"Yes," said Lee. "They are alive but unconscious."

"I should be treating them," said Zakany. "I'm the doctor."

"They're in good hands," said Lee. "Soon, you'll back up and bossing everyone around."

Just then Lieutenant Andras showed up with the rest of his team. He reported to Aruban. "Sir, all accounted for. All have minor injuries; one is more serious and will require a doctor when we return."

"What happened?" asked Aruban.

"We attacked the flank, as directed," said Andras. "We pulled in lots of Nephilim; way more than we could handle. We would have had a hard time of it, except that the diggers rose up. They were awesome. They even gave us an escort coming here. I take it they are our friends now?"

"I suppose," said Aruban. "Prioress Sarnai has taken them under her wing."

Lee walked over to Aruban and Andras. "What happened at the village? Will we be needed at the wall?"

Andras shook his head. "No. It's the same there as here. The Olgoi rose up and attacked the Nephilim. None of the giants survived. Some of the smaller ones ran away. But the village is secure."

CHAPTER 47

THE AFTERMATH OF BATTLE

As Lee reentered the walled village at dawn, she saw the devastation. Both of the external walls were severely damaged, and in some places had collapsed due to fire. There were bodies and parts of bodies everywhere. Both attackers and defenders had put up a furious fight. There was no way to tell what portion of the damage had been done before or after the Olgoi intervention.

No stranger group had ever entered the village. Sarnai and Andras led, followed by the 12 scouts, who were carrying five stretchers between them. One stretcher carried a scout who had been severely injured, and the remaining four carried the shuttle crew, all of whom were either still unconscious or too weak to move on their own.

Lee walked alongside her crew, speaking softly and occasionally reaching to give a reassuring touch. She knew the big risk was shock for those having been tortured and denied food and water over a long period of time. A friendly voice and touch could prevent its onset.

Next came the Olgoi-Khorhoi, five of them led by Arslan. Lee was once again surprised to see them all walking on their hind legs. They actually looked quite natural, if not elegant. The remainder, some 50 diggers, stayed outside the city walls. They had very quickly created dirt walls about two meters high around the entire wall structure and set up what looked like a military camp inside it.

When Lee asked her about it, Arslan had said, "For now, until the exterior walls are rebuilt, Olgoi will protect the city. They are also constructing living quarters underground."

Once inside the walls of the village, Lee could see that at some point the Nephilim had broken through into the village. There was evidence of a horrific fight that had taken place in the first few blocks. Many of the buildings had been burned or had collapsed after having been damaged.

At the farthest limit of the damage, it was clear the fight had been the fiercest. She saw the bodies of three Nephilim, or at least she thought it was three. It was hard to tell because the bodies were in pieces. She assumed the Olgoi had ripped the Nephilim apart.

The villagers themselves surprised her. There was no wailing or crying because of the death and destruction wrought upon them. They were steadily clearing away the debris and carefully recovering the bodies of their own loved ones.

It amazed her that the villagers showed no surprise at the presence of Olgoi accompanying the procession. There were a few looks and even fewer double takes. Lee supposed the villagers had already seen the Olgoi turn on their masters and fight for the village. Perhaps everything after that would seem mundane.

Lee saw the child, Bayalag, whom she had spoken with briefly the night before. The child was covered in dirt and soot. She stood alone by the side of the street, eyes cast down. When she looked up and saw Lee, she brightened and shouted. "Warrior Goddess, you saved us!"

Lee smiled and waved a friendly gesture. At this exchange, everyone within hearing distance stopped what they were doing and moved to the sidewalk and began clapping and cheering. Lee heard the phrases: *Daichi Tengri, warrior goddess, prophecy,* and finally, *killed five Nephilim with her own hands.*

CHAPTER 48

DAICHI TENGRI OR NOT

Once back at the monastery, things began to fall into an order that Lee found reassuring. The injured went to the infirmary. Five patients were considered critical: Lee's four crew members and a scout who had a serious injury to his leg. The monastery had an impressive medical capacity. Lee didn't recognize the ointments and herbs being used, but she saw no reason to object. She didn't have anything better to offer and of course did not have access to an autodoc.

Within a day, Laura was back on her feet and helping with the care of the wounded. Lee asked her about the medical care they were getting from the monastery.

"It's mostly a homeopathic approach," Laura said. "It's not harmful and some of the medicines they are using seem to be helpful at preventing infection. Luckily, what the four of us need most is rest and hydration."

"You had a rough time," said Lee. "Do you want to talk about it?"

"Of course," said, Zakany. "We were lucky when the shuttle disintegrated, and we ejected. We all wound up in pretty much the same area, all within about a kilometer of each other once on the ground." She smiled and continued. "By the way, that was some ride on the way down. I've never done a parachute drop. I might like to try it again under better circumstances. Will you show me?"

Lee smiled and said, "That would be fun. My chute malfunctioned and I ended up using the wing suit and pretty much crashing. The monks found me and brought me here."

They were silent for a long moment. Lee understood. Laura would say what she needed when she was ready.

"The smaller ones found us first," said Zakany. "In the beginning, we thought they would help us, but they didn't. They are fast and smart. They took us quickly before we could get out of the parachute rigging. Even so, we put up a good fight. Ali got to his sidearm and took out three before he was overwhelmed. The rest of us were taken too quickly."

"They hurt you," said Lee. It was a statement, not a question.

Laura's faced went blank, colored and then she said, "Yes, they bound us, as you saw, and tortured us. They didn't seem to want anything other than the pleasure of hurting us."

"Do you know why they brought you here?" asked Lee.

"No, but they seemed intent on it," said Laura. "It was as if our arrival and capture was the trigger for the assault on the village. They brought us through this system of tunnels. It was terrifying, completely dark, stuffy and claustrophobic."

"Do you know anything else about them? Did they say anything?"

Laura considered, "They do have a language they use, but I couldn't understand what they said, just a few snippets. They kept repeating something like *dotty ten greer.*"

Lee was astonished, "Was it *Daichi Tengri?*"

"Yes, it could have been," replied Laura. "What does it mean? I heard the villagers saying it as we were being brought to the monastery."

"Daichi Tengri is a mythical warrior goddess sacred to the Mongolian people. For some reason, they think I am her, or a reincarnated version of her."

"Didn't you tell them you're not?" Laura asked.

"Yes, of course," said Lee. "But their take on it is that I could be the reincarnated version and not know it myself. So, my denial doesn't rule it out."

"Why you and not someone else?" Laura asked.

"They apparently have a thousand-year-old prophecy that Daichi Tengri would return someday, riding black wings from the sky, and that she would help them at their hour of greatest need."

Laura looked at her quizzically.

"There's more," said Lee. "They have a 1,200-year-old tapestry that shows the goddess coming from the sky on dark wings. It's part of the prophecy. The image does look uncannily like me. Their own legends have it that Daichi Tengri brought their ancestors here in the year 1200 CE on some sort of space craft."

"But that was well before FTL became available," said Zakany.

"Yes, and that's a problem," said Lee. "I think they *have* been here that long. They got here somehow, and I have no idea how that could have been possible."

Laura looked at her old friend, cocked her head and said, "There's more isn't there?"

Lee nodded. "Yes. Its sounds a bit crazy. Remember when I was trying to deal with the warhead during the EVA?"

Laura nodded.

"It was very odd; I heard voices, saw faces. But when I touched the warhead, I felt as if I had been transported to another place. I was on my back, looking up and a saw a face peer over at me saying, "There you are.""

"OK," said Zakany.

"When I landed here, after crashing into a tree line and hitting an open meadow, one of the monks, Guyuk, was there. He said the words, '*There you are.*' It was just like in the vision."

"Well, that's different," said Zakany.

"It actually gets weirder," said Lee. "Guyuk had had the vision of me crashing in that meadow some 30 years prior. He told the Abbot about it at the time. The Abbot believed Guyuk foresaw the fulfillment of the prophecy of the return of Daichi Tengri. Guyuk was actually in the meadow waiting for me, as he had every day for 30 years."

Zakany nodded. "I understand. You and he both had the same vision. His was 30 years ago and yours was a few hours prior."

"You don't seem surprised," said Lee.

Zakany shrugged. "*It happened,* so there's no sense in being surprised. Life is strange; we don't always get a chance to understand even simple coincidences. This is bigger. Something is happening, and you, all of us, are a part of it."

"What do you think it means?" asked Lee.

"It doesn't have to mean anything," replied Zakany. "It is whatever it is. Doesn't really change anything, does it? You will still need to do whatever it is you need to do, Daichi Tengri or not."

CHAPTER 49

THE PLACE OF SOULS

Lee was resting in her quarters when a knock came at her door.

"Come in, please, it is not locked."

To her surprise, Arslan, the chief of the Olgoi-Khorhoi entered standing on her hind legs and bowed deeply. Lee had been told that Arslan's title was actually speaker, not chief. She wondered what the distinction was.

"Madam Speaker," said Lee. "It is an honor to see you. How can I be of assistance?"

"Daichi Tengri," she said in her rough speech. "The honor is mine."

"Please, come to the table and sit with me," Lee motioned to a small table near her bed that had two chairs. "Can I call for refreshments?" Lee didn't know what the Olgoi ate and so was unsure what to offer.

Arslan made what was clearly a smile and said, "If it is not too much to ask, I would love coffee. That is a rare treat we have not had in a long time."

"Excellent," said Lee. "I have a thermos here with coffee made from Marine rations."

Lee got out the flask and poured them both a cup of the steaming liquid. They both took a sip and Lee could see that Arslan relished the drink.

As Lee waited politely for Arslan to get to business, she discreetly studied her visitor. Intelligent non-human species were rare, at least as far as the Alliance had encountered during its expansion into the galaxy. Arslan was covered in fur that had a somewhat luxurious look to it, mostly grey with an almost blonde-brown color in other parts. She had facial features that, to Lee, looked like something between a fox and a wolf, except the jaw was not so pronounced. The eyes were yellow, clearly intelligent, expressive and thoughtful. She gave off a not-unpleasant scent that reminded Lee of incense and sandalwood.

Finally, Arslan put down her cup and looked Lee directly in the eye. It was an intense and disconcerting gaze. No human could actually have that effect, thought Lee.

"On behalf of my people, those you call the Olgoi-Khorhoi, I thank you for saving us once again," said Arslan.

Lee bowed her head slightly to acknowledge the gratitude. She said, "You are kind, Speaker, to give me such credit. If not for the bravery of the Olgoi, I fear all of us would have been lost to the Nephilim."

"Had you not returned, we would have remained in bondage," said Arslan.

"Madam Speaker," said Lee. "I must be honest with you. I do not believe I am this Daichi Tengri. I am 30 years old. I was born on Earth, and prior to coming here by accident a short time ago, I have never been to this planet, nor did I even know it existed."

Arslan met this confession with a slight smile. She visibly took a deep breath through her nostrils, let it out, and spoke. "Did you know that my kind have a very-well developed sense of smell?"

Lee said nothing, and Arslan continued. "For example, I can tell when someone is being truthful or not, just from the scent one gives off when speaking. For us it is quite pronounced and unmistakable."

"So you know I am telling you the truth when I say I am not Daichi Tengri," said Lee.

"I know you believe that you are not the Goddess of War we call Daichi Tengri," said Arslan. "But I also know that you have some doubts and confusion. You do have long-buried memories that you cannot explain away."

"You are insightful," said Lee.

Arslan took another deep breath, paused and closed her eyes. After a moment, her eyes snapped open with a look of surprise and she said, "You have been to the place of souls." It was a statement, not a question.

Lee waited to gather her thoughts and replied, "When we were on our way here, I had to go outside the vessel under unusual circumstances. While I was out there, I saw faces and heard voices. I thought it was just a hallucination. But one of our automated devices sensed them, and even tried to fire on them."

Arslan remained silent. At length, Lee said, "You called it the place of souls. What is it? Have you seen them, too? What are they?"

"They are what you suspect, the souls of those lost, who cannot find their way to a place of peace."

"They hammered at my face plate," said Lee. "They were saying things, even screaming, I couldn't understand it."

"They are lost, and desperately seek a way home, or at least some relief."

Both were quiet for a while. Then Arslan spoke. "My species, we communicate through scents, not just spoken words. One thing we are taught, all of us from birth onward, is to recognize the scent of the Daichi Tengri. You have that scent, there is no question, no doubt. You are not some reincarnated spirit. You are her."

Lee sat and thought for a long moment. Finally, she said, "I respect your views and your conviction, Speaker. But I must rely upon my own memories and the known, verifiable facts of my life to embrace my own identity."

"That is as it should be," said Arslan. "But know this. The Olgoi believe that Daichi Tengri will someday free the souls of the lost. Your experience confirms my understanding of who you are."

"Speaker, you said that I had saved you *once again,*" *said Lee.* "Can you tell me in what other instance such a thing has happened?"

Arslan nodded and said, "We, the Olgoi, are the original inhabitants of this planet. For millennia we were the only intelligent species here. We built our cities beneath the soil and lived in relative harmony with each other and with the natural order." She paused and looked again intensely, perhaps questioning Lee.

"That is commendable," said Lee.

"Many hundreds of years ago, humans like you, came here in a vessel that they said sailed both time and space," said Arslan. "They were not good people, unlike you and your colleagues. We, of course, could tell through their scent that they meant harm to us, so we resisted their coming. There was a battle, and the Olgoi lost. Those who survived were enslaved and forced to dig for the bad humans."

"What was it they forced you to do?" asked Lee.

"They made us bury a substance, a type of rock. They wanted it deep in the planet's crust, very deep. It was a difficult and dangerous task. Many died."

"And your people did this?" Lee asked. "How did they get free?"

"Our legends tell us that our leader, our Speaker at the time, Ganzorig, escaped from the craft upon which he was being held as a hostage along with a another. He was being chased and would have been killed but you, Daichi Tengri, came to his aid. You called him by name. You fought the bad humans, killed them or drove them off, and freed the Olgoi."

Lee said, "The Mongols here also have a story that I brought their ancestors here from earth over a millennium ago."

"Yes, that is true," said Arslan. "Our ancestors were aware of it at the time."

"And did you have peaceful relations with them, the Mongolians?"

"Yes, they are a peaceful people. We learned their language, which is very difficult for us because we use our scents to convey context and meaning. Humans do not use scent in the same way. We find your language very limited, like speaking with a small child. There was some trade between us early on. But eventually, they stayed in their places and did not seek us out or try to harm us. Over the centuries since, we have had little, if any, contact."

"What of the Nephilim?" said Lee. "Are they also native to this planet?

"Yes and no," said Arslan. "Believe it or not, they are related to us, the Olgoi. They were a subspecies of the Olgoi. Not intelligent, just one of the many species on this planet. They lived in the forests. We had almost no contact with them, and certainly no conflict. About 800 years ago, they somehow evolved quickly. They grew larger, fierce, violent, and semi-intelligent. We don't know why, but their scent changed. That is something that is impossible in nature. We believe that they were interfered with, and in some way altered. We tried to make peace with them, but they attacked us and eventually enslaved us."

"Why did they attack the village? Why take my friends hostage? Why inflict such cruelty?" asked Lee.

"They came for you, of course, for Daichi Tengri."

"I don't understand," said Lee.

"Nor do we," replied Arslan. "When your ship crashed and your colleagues were captured, it triggered something urgent and compelling in their behavior. They kept repeating the words, *Daichi Tengri.*"

"For what possible motive?" asked Lee.

"We don't know for sure, but our best guess is that by holding your friends hostage as they did, they were trying to get you to come to them."

"That makes no sense," said Lee.

"No, it doesn't. Not to us anyway," said Arslan. "But recall, they are only semi-intelligent. Their culture is based on extreme violence. It may have been their only way to try and reach you."

"And why would they want to?" Lee asked. "I have nothing to do with them."

Arslan shrugged and said, "There are many mysteries here."

Arslan paused as if considering her next words. "But consider this possibility: The Nephilim capture your colleagues. Like the Olgoi, the Nephilim have a well-developed sense of smell. Assume they pick up your scent — a person they recognize as the War Goddess — Daichi Tengri. That smell is on the clothes of your friends. In the minds of the Nephilim, your friends, who are all armed, have harmed or captured their goddess. They lay siege to the village and crucify your colleagues as a way to compel the villagers to give you up."

"Do you think that's what happened?" Lee asked. "That they were trying to save me?"

"That might be a partial truth," said Arslan. "The overall reality is likely to be more complicated than that."

"They tracked me by my scent, then they attacked me," Lee said simply.

"I know, I was there," said Arslan. "But know this: when in combat, the Olgoi system of scent is refocused for fighting. We lose all nuance. Recall I didn't recognize your scent until I took your hand. The Nephilim's ability to identity you may be the same. You were with Prioress Sarnai. That may have confused the scent. They were in combat; they may not have recognized you."

Lee froze with a sudden memory.

"What is it?" asked Arslan.

187

Lee looked directly at Arslan and said, "When I killed the last Nephilim, it was close combat. I had to get on its back and use my knife. At the last moment, it stopped struggling, let out what sounded like a mournful cry. Could it have recognized me then? Maybe it was trying to surrender?"

They both sat in silence with their own thoughts.

Then Arslan said, "We knew you would return, Daichi Tengri, because you told my ancestors that you would. And here you are."

"Why tell me all this, Speaker?" said Lee.

"I tell you this because there is a great struggle ahead for you and for this planet."

CHAPTER 50

A MEDICAL ENIGMA

In the weeks that followed, the work of repairing the village and the defensive walls took place with a sense of urgency. No one could be sure that the Nephilim would not return, perhaps next time in greater numbers.

Aruban sent out patrols to seek out Nephilim that may have escaped, but they could find nothing. The Olgoi returned to the tunnels they had created when they were slaves to the Nephilim, and filled them in.

Danner, Ali and Odessa regained partial consciousness after about a week, but something was wrong with their recovery.

"They show a profound malaise, some sort of depression," said Zakany to Lee over breakfast in the monastery refectory. "They are awake only a few hours a day, and when they are awake, they are glassy-eyed and unresponsive."

"What's causing it?" asked Lee.

"It's called post-coma unresponsive state or PCU. That's what their symptoms show anyway," said Zakany. "It's a well-known condition following a coma, but..."

"But what."

"It's a very rare condition," said Zakany. "Only about four percent of patients in a coma display these symptoms coming out of a coma."

"So, they have a rare condition," said Lee. "Why the '*but*'?"

"To have all three with the same condition is extremely unlikely. I don't have an autodoc or any other medical equipment I could use to do a diagnosis."

"What does Sarnai say about it?" asked Lee.

"She thinks it's from a pathogen of some sort," said Zakany. "She notes the three of them have bite marks from the Nephilim which, fortunately, I do not. Sarnai says a coma like this is an expected result from Nephilim bites."

"So, what's the treatment?"

"Sarnai has a regimen of traditional Chinese medicine: incense, herbal rubs, they even use a reed tube to blow smoke into the ears. None of it is harmful, so I haven't objected. And who knows, maybe it will help."

"Prognosis?" asked Lee.

"If these are PCU cases, prognosis is poor," said Zakany. "Almost all patients with these symptoms stay like that forever or slip into a persistent vegetative state which is worse. That's where there is no visible sleep cycle and no responsiveness to stimuli."

"And if it's not PCU? If it's something from the bites?"

"I just don't know," said Laura. "We are all inoculated against all known pathogens, but this is something outside our known medical science."

"So, we wait?", asked Lee.

"Yes. Sarnai's team is keeping them comfortable," replied Zakany. "Their aroma therapy won't hurt and it might help. Best case scenario, we are looking at weeks or months for improvement."

"Worst case?" asked Lee.

"Worst case: permanent PCU, or degrading into a persistent vegetative state, followed by death."

"Death...," said Lee. She looked away and breathed deep. Lee thought about all the people she had lost over the past few years.

"What will you do?" asked Lee.

"I'll keep working in the ward here," said Zakany. "There are dozens of wounded from the fight with the Nephilim. I can be of help. Sarnai and her medical people are open to some of what I have to show them, and I'm learning from them as well."

They were quiet for a moment, each in their own thoughts. Finally, Laura asked, "What's the status for getting us out of here?"

Lee took a deep breath and let it out. "I don't know. If the pilots were able to get a distress signal out before we entered the atmosphere, then someone in the Alliance will come for us eventually. But even if no signal went out, we are overdue and missing. A search should be undertaken."

"That's good, or at least hopeful," said Laura.

Lee shook her head, and Laura asked, "What?"

Lee said carefully, "The problem is that I don't know how much of our mission was classified, or how deeply classified. Danner had the lead. It was his mission, and he can't tell us. Recall that not even the security team he had with him was allowed to come with us."

They were both silent for several minutes. Laura finally said, without looking up from the table, "I have a two-year-old son. I really do need to get back."

CHAPTER 51

YOU HAVE FORGOTTEN

Sarnai and Lee stood on a balcony overlooking the training area, which was a large courtyard about 3,000 feet square. Lee had asked Prioress Sarnai for permission to observe the monks' physical training. Sarnai had agreed.

Lee had been taught a traditional version of Northern-style Kung Fu until she was about eight years old. At that point, her father had trained her in his own style of martial arts called Aristos. Aristos had incorporated many of the techniques of different martial arts but applied a philosophy of classical Greece, rather than the Confucian ideals to which many martial arts paid only lip service.

In the courtyard, about two dozen men and women were training. They were performing what Lee assumed was a variant of one of the forms of Northern Kung Fu. It was different than the way Lee had learned it under her father's instruction. The form, Kaimen, which meant *Open the Door*, was taught to her as almost a form of dance, with many flowing movements, and long extended stances.

The techniques the monks were performing, however, were stronger, more direct. The stances were solid, designed for stability and power.

As she watched one of the monks closely, something seemed wrong, disorienting. It was as if his movements were disjointed, like a glitch in a video display.

Lee turned to Sarnai and said, "I see women train with the monks."

"Yes, women are trained from childhood, as I was," said Sarnai.

"Are you nuns, then?" said Lee.

"No," she responded. "The women are considered equal to the monks. We are addressed as *sisters*, but we do not take vows, as some of the monks do. We are free to marry if we choose.

Seeing Lee on the balcony, Guyuk, the same monk who had found her in the meadow after she crashed, called up to her and asked if she wished to join the training. Since she had no other duties as long as her team was recovering, she had agreed.

Now, two weeks later, Lee trembled with exhaustion. She had been holding a modified cat stance for several minutes. Guyuk, ever cheerful and respectful, could be a relentless instructor. Not since she left home for the Naval Academy had she been so tired from training in the martial arts.

On the day she joined the training, Guyuk had asked her to show him her forms and basic techniques.

He beamed. "Oh, that is very good. Very strong. May I suggest a modification?"

Guyuk had proceeded to show her, with utmost courtesy and good humor, that all her techniques were incorrect, or at least incomplete.

At his request, she had demonstrated various kicks, including a high side kick designed to strike an opponent in the face.

"Oh, my, that is impressive," he beamed. "But I fail to see the purpose. Why would you risk an upset, when you could attack the legs or torso so much more safely?"

Lee did not have an answer and so said nothing. Guyuk was right, of course. High kicks looked good, but they came with a risk. A person executing a high kick would be slightly off balance. An experienced defender could trap the kicking leg and sweep the supporting leg.

Which is exactly what he did when they first sparred. Guyuk had a habit of making a clicking sound with his tongue. She found it distracting and forced herself to try to ignore it. He also moved his hands up and down and side to side. It was an odd movement, and she couldn't understand what advantage it brought him. To her it looked like an unnecessary, quirky movement.

She kicked at his head with a front leg roundhouse, and he easily stepped inside the kick, caught her leg in the crook of his arm, and swept the supporting leg. Down she tumbled onto the dirt floor of the training area.

But there was something odd about the exchange. First, there was no way he should have been able to close the distance to her at the speed he must have traveled to make the takedown. Lee was a very experienced martial artist. She had trained to the master level while still a teen and had been a champion competitor at the academy. Additionally, she fought in combat many times in the past two years and used her martial arts skill to good effect.

How had this tiny, elderly monk beaten her so quickly and thoroughly? She replayed the exchange in her mind. Her kick itself was fast, only a fraction of a second. But before it could land, he was already slamming into her, and sweeping her other leg. It was as if she had lost a moment in time, because she could not recall him moving at all. One second, he was her target, the next he was in her face.

That was two weeks ago. She had spent hours a day under his tutelage training basics of stances, blocks, parries, movements, strikes and kicks. Over those days, some of his approach became clear.

First, the stances were made to be strong. Somehow, some way, once in a defensive stance, it was as if he was cemented to the earth. He

194

could not be thrown, nor even moved. At the same time, the stances allowed him to move more rapidly than she would have thought possible. As a result, he could choose to accept an attack, stay put and block, or trap the strike. When he did this, he was immovable.

Or he could dodge any strike if he so chose. Lee had a humiliating session of sparring in which Guyuk simply weaved and dodged all her attacks, sometimes laughing in simple delight at his game. She found herself literally punching the air.

She tried to figure out how he was doing it. His footwork was exceptional, that was certain. But there was something else. He seemed to know in advance what she would do, and so he was never surprised. Lee had been trained to *never telegraph* her moves in advance. In fact, it was a core tenant of her art to give no advance warning of any move she might be anticipating. No shifting of weight, no movement of the eyes to give away intent, no tensing of muscles to indicate her next move. But he seemed to know. *How was he doing it?*

There was something else, something far more troubling. At times she seemed to lose track of where he was. It only lasted a moment, but it was uncanny, and it threw her off of any tactic she might be using in offense or defense.

Frustrated, Lee gave into her anger and began to spar in earnest. She would show him her best. Why not? He could take it, surely. As she began to go onto the offense will full power and speed, Guyuk noticed and said in a lighthearted voice, "Oh, my. Such forcefulness."

At first, he seemed to give way. She felt at least she was forcing him to retreat, if only a little. This minor victory caused her to increase the ferocity of her attacks. Guyuk lost the smile, which she interpreted as though he was finally taking her seriously as a fellow martial artist, rather than an awkward student.

And then all of a sudden, he was in her face, his wide friendly smile returned, and he very gently but firmly placed his forefinger on her nose and said wryly, "You're it." In a flash he was gone again, laughing lightly, as she futilely tried to reestablish dominance.

Lee was flabbergasted. *Just not possible.* How had he penetrated her defenses, getting in so close that he could touch her, and then slipping away without her able to lay a hand on him? No one she had ever trained with or competed against could do that, not even her father who was one of the most prominent martial artists on Earth.

She went hard at him again, and time after time he got inside her defenses and touched her with his finger and then he slipped away.

Finally, she accepted this was happening. She stopped, bowed and said, "Master monk, I would be very grateful if you could show me how you are doing that."

"Yes, yes, *Daichi Tengri*," said the monk. "You have forgotten that *you* showed my ancestors these things many centuries ago. You have forgotten, but we never will. I will show you, if you will allow it."

"I allow it," replied Lee.

CHAPTER 52

SOUL SLIDING

Under Guyuk's tutelage, Lee spent the next month in the hardest training of her life. Physically, she was pressed to her limit. She moved into a room off the training courtyard. It was eight by ten feet, with a single stone bed and a thin blanket, a toilet and a sink. As sparse as it was, she was glad to see that bed at the end of a training day that sometimes ran 20 hours long.

She took her meals with the monks and sisters in the refectory. She was accepted as if she were a member of the order, the monks and sisters were friendly, but no longer deferential. Meals were very basic, but nourishing. She found herself looking forward to this time, partly because the food was good, but mostly because it was a time of respite from Guyuk's relentless training.

He explained to her what she needed to do.

"You must unlearn your very bad habits," he said. 'There is your body, your intellect, and your spirit. In you, these are all strong, but they are separate. They do not move together. When we spar, I see you and you do not see me."

Guyuk's approach was to wear her down physically to the point of collapse, and then build back her ability to *see,* as he put it. Lee spent long hours doing calisthenics, strength training, running, martial arts forms and sparring. Guyuk would only let her spar when she was near total collapse from exhaustion. She would be matched with one of the younger monks or sisters in training.

He would say, "You are not watching. See your opponent. *See him!*"

Lee had no idea what he meant, so she just tried concentrating. For weeks nothing happened.

Then something did. She was sparring a young monk about 20 years old. As she looked at him, she saw a vague shimmer, almost an outline of his body. At first, she thought it was a hallucination brought about by her exhaustion.

As she concentrated, she saw the shimmer take on the shape of the monk. Then something odd happened. The ghost-like outline moved separately from the monk, and it moved to her right, launching a kick to her thigh. A fraction of a second later the monk then executed that very move.

What the hell? thought Lee. She was so surprised; she made no defense, and thus the kick landed painfully on her upper thigh, causing her to collapse from the force of the blow and the accompanying pain.

She lay on her back trying to digest what had just happened. Then she saw above her the smiling face of Guyuk. "You saw it, yes?"

"I saw something, a shadow of the man," said Lee. "It moved separately from him, then he followed and did the very same thing. What is it? What did I see?"

"If you will get to your feet," said Guyuk. "I will try to explain."

Lee struggled to her feet, shifting her weight onto her uninjured leg. She saw the young man, Batu, whom she had been sparring, kneeling face cast down.

She spoke to him gently, "Master Batu, please rise. There was no foul; the fault is mine."

Batu nodded and rose to his feet quickly and bowed politely.

She turned to Guyuk with a questioning look.

He said, "You saw his aura, his *hun*, maybe his soul. It has its own intention, and it moves separately from the person. Every person's hun is different, quite distinct and identifiable."

"But it moved before him, and then he followed," said Lee.

"Yes, that is correct," said Guyuk. "The hun sometimes does that, it is not always so, however. One must be careful not to rely upon it all the time when fighting."

"And you see mine?" said Mei Ling. "My hun, when we spar?"

"Yours is very strong, not like anyone else's," said the monk.

"And you use my *hun* to predict what I will do," said Lee. "In order to get inside my defenses?"

"In part, yes," he said. "We call it *soul sliding.*"

"What else?" said Lee now feeling a little defensive.

"Even without the appreciation of your aura, you are clumsy and slow," he said without any indication that he was aware he might be insulting her. Lee was not angry. In fact, she smiled. Here was a man who said what he thought. And that, to her, was refreshing.

Guyuk shrugged. "To be fair, I have other tricks I used on you. Totally unfair, I admit. Would you like me to show you?"

Lee nodded and was about to say, *"Please show me,"* when Guyuk was right in her face once again, touching his index finger to her nose.

"What the hell!" exclaimed Lee. "How did you do that? I didn't see you coming."

CHAPTER 53

SLEEP MAKING

Guyuk smiled at her in a conspiratorial way and clicked his tongue and moved his hands up and down, side to side.

"What?" asked Lee.

"You watched the hands, listened to the clicks," he said. "Then you go to sleep, then I move."

"You're saying you are hypnotizing me?" asked Lee in astonishment.

Guyuk shrugged and said, "I don't know the word you used. We call it *sleep making*."

"How does it work?" asked Lee.

"I start by making the clicking sound. I watch how you respond to it, then I change the cadence, and move the hands so you are watching and listening. Even if you try not to, you unconsciously follow the rhythms of the clicks and the movement of my hands. Then I change it up, just a tiny amount. You look for the old pattern, seeking harmony, but it is not there. When I see the slacked look in your face, I know you are elsewhere. Then I move in to say *hello*." Guyuk gave a delighted smile.

"Can you teach me?" asked Lee.

"Can I teach a 30-year-old adult such a trick?" asked Guyuk. "No. One must begin to learn as a child, or else it will not take."

Lee frowned, and Guyuk continued. "But you do not need to learn from the beginning. You know this technique. You taught it to the monks when you brought them here so long ago."

Lee was frustrated and said, "Why do you insist I am this *Daichi Tengri?* I'm not. I'm Mei Ling Lee. I was born in Hong Kong thirty years ago. I'm not a goddess and, until I crashed here six weeks ago, I didn't know this planet even existed."

Guyuk shook his head. "Some things don't lie. Everyone who can see an aura recognizes you as Daichi Tengri. The Olgoi elder, Arslan, recognized you by your scent. Even the Nephilim knew you. There is no mistake. You are Daichi Tengri."

Lee threw up her hands up in frustration and said, "OK, I give up. Please teach me if you can."

"It will be my pleasure, Daichi Tengri," replied the ever-smiling monk.

CHAPTER 54

DARK TIMES AHEAD

After another month of, Guyuk changed up her training. At the beginning of the training day, he had her put on a specialized vest that pulled her left arm close to her torso and held it there tightly. He required her to go through the entire training day with his contraption, and he made no concession for it in her training. A first she found it very awkward. She had never realized how much she depended on her arms for balance, and she found herself falling down, sometimes skidding on her face.

After a few days of this, he fitted her with another vest, this time restricting her right arm. Again, she went through a period of adjustment before she could maintain her balance.

The hardest part was sparring. She had to spar the other monks-in-training, and they, of course, did not have any such restrictive clothing. Guyuk ordered them to attack with full force, and even to try to take advantage of Lee's one-arm status. Many times, she got hammered and was knocked to the ground. Guyuk gave her no help of any kind. No techniques to use when fighting one-handed. He just ruthlessly threw her into the match after match.

Eventually, and out of necessity, she learned ways to cope with the impairment. Her balance improved, and with that, her options to move, block and attack became evident.

Just when she felt she had reached an acceptable level of proficiency in this type of impaired fighting, Guyuk added a twist: more opponents. Again, Lee found herself on her back as the two monks would rush her kicking and punching with full force. Slowly, ever so slowly, she learned to use the *Soul Sliding* technique to anticipate their movements, and either avoid them or attack their vulnerabilities. Once she even defeated two at the same time, knocking both to the ground hard enough that they didn't get up right away as they usually would.

After a particularly long, brutal day of this, Lee asked Guyuk, "Why all this sparring with one arm?"

Guyuk looked sad for the first time she had known him. He looked her directly in her eyes and said, "Daichi Tengri, there are dark times ahead. I see treachery in you your path. It is best to be strong."

And he would say no more.

CHAPTER 55

DANNER WAKES UP

The month stretched into another and then another. Lee checked on the three unconscious crew members every day. She checked in with Zakany on their progress and sat with the unconscious ones. Touching their arms, speaking softly to them. But there was no change.

She spent her mornings training with Guyuk and some of the other monks, perfecting her technique and learning, ever so slowly, the secrets they held. After a light noon meal, she would strike out into the surrounding hills for long hikes. It was a time of deep breathing, meditation and serenity. She loved the woods and paths she explored. Occasionally, Sarnai would accompany her, and Lee would ask her to instruct her about the trees and other plants along the way.

Lee had been born and raised in Hong Kong, and so she was truly a city girl. Her only exposure to nature had been in the Marine Corps, and that had mostly been just enough understanding to exploit terrain for tactical advantage.

Then one morning as she was sitting next to Danner, talking to his unconscious form and patting his arm, she said, "I was out walking yesterday, and I saw this huge bird. It looked like a mixture between a dodo bird and an iguana. Sarnai says it's called a *Pinyin*."

Danner opened his eyes and said quite conversationally, "Isn't Pinyin the Mandarin name for a Crane?"

Lee started and jumped up. "Danner! Are you awake?"

"Pretty sure I am," he said and cracked a weak smile.

"Laura!" Lee shouted. "Danner is awake!"

Laura came over and smiled broadly. "Well, I'll be. Sleeping Beauty has awakened." She winked at Lee and said facetiously, "I hope you didn't kiss him."

Laura's expression turned serious, and she sat down next to Danner. "Let's have a look."

She checked his pulse, his blood pressure, shined a light into his eyes, made him open his mouth, and asked him a battery of questions to check for his overall cognitive ability. He answered all questions slowly, as if he had to think about it; but all were correct.

"Now I have some questions," said Danner. "What's happened to me, where are the two pilots, and where are we?"

Zakany filled him in. "After we ejected, we were taken prisoner by a race the locals call the Nephilim. We were tortured, brought to the outskirts of this city, where Mei Ling was already in residence. The Nephilim attacked the city. Mei Ling and the local constabulary rescued us and, with the help of another species, drove off the Nephilim. You, Ali and Odessa were bitten by the Nephilim. However, I was not. We believe there is a venom in their bites that triggered a coma. The three of you have been in a semi-coma for months. You are the first to come fully out of it."

Danner didn't seem surprised by any of it. He just nodded and took it all in. He looked at Zakany and said, "Thank you for taking care of me."

Then he looked at Lee and said, "Thank you for coming to get us. I remember some of it." He looked away and said, "Some if it was very bad."

Then he fell back asleep.

CHAPTER 56

STOCKTAKING

It took a week for Danner to recover enough to seem, at least in part, more like his old self. At first, he couldn't walk on his own and he tired easily. But Sarnai and the medical personnel had experience in bringing bite victims back to reality.

As Lee watched, she could see that the rehab was hard for him. The experience had left him with some nerve damage, and the months of immobility had left him weak. More than that, he seemed depressed. His expression was often slack, and he didn't always meet her eyes when they spoke. He spoke in a near monotone, and would answer questions put to him, but rarely initiated conversation.

Two weeks after he woke from his coma, Lee and Zakany sat him down for a discussion on the way ahead.

"Andy," said Lee. "Are you well enough to talk about operational matters?"

Danner looked up and said simply, "Yes."

"First, please tell us what the mission is," said Lee.

Danner nodded as if remembering and said in that deadpan voice, "The mission was to make contact with Admiral Chambers and Warrant Officer Nemeth."

"So, they are alive," said Lee. It was a statement, not a question.

Lee turned to Zakany and asked, "Did you know?"

"Yes," Laura replied. "Danner told me that Mathias was alive, and that I was needed on this mission. I don't know anything beyond that."

"Background?" asked Lee.

"Our intelligence agencies monitor, or try to monitor, systems where the Alliance no longer has a naval presence," said Danner. "About 18 months ago we got word that a previously unknown warlord by the name of Mr. Watson had seized control of a planet and some asteroid mining operations in the TOI-700 system. By the description of the man and his deputy, we believe this person is Admiral Chambers. You will, of course, know that Watson is his middle name."

"What do you think he is up to?" asked Lee.

"We don't know a lot of things. We don't know why Admiral Chambers disappeared after the Battle of Alpha 51; we don't know why he seems to have taken over a criminal enterprise."

"They must have a good reason," said Zakany forcefully. "Mathias would not act dishonorably, and he would not abandon me and our child."

"I agree with Laura," said Lee. "Chambers and Nemeth are two of the most honorable people I've ever known. If he has jumped into the criminal world, he has a reason, and it's not profit."

"What does Naval Intelligence think?" asked Lee.

"They don't know Chambers and Nemeth like we do. And however unlikely it is for them to have gone rogue for profit, we've seen far worse with other high-ranking people conspiring against the Alliance, attacking our cities."

"What more do we know?" asked Lee.

"There are some oddities," said Danner. "First, the planet he picked is quite specific. D-4, or Ulysses as it is now called, is in the TOI-

700 system. It is well out of the way, as it is 100 light years from Earth. He needed an FTL drive to have gotten there and we don't know where he got that; the shuttle he was in when the two of them went missing didn't have it. He either acquired a different vessel or he modified the one he had."

"What's out there in the system that might have drawn his attention?" asked Lee.

"Before Chambers and Nemeth arrived, a criminal gang took over the planet, and began mining the Oort Cloud of that system for Boron Nitride, which is one of a handful of materials needed for quantum computing," said Danner.

"Is this material being sold on the black market?" asked Lee. "I assume it's very valuable."

"You're right, it is extremely valuable," answered Danner. "And no, none of it has shown up on the black market. Whatever they're doing with it, they're not selling it."

"How did Chambers and Nemeth muscle in on a criminal enterprise?" asked Lee.

"We don't know the details," said Danner. "They were both formerly in Oracle; serious special operations warriors. It looks like they may have posed as members of a rival gang and come in very hard. Lots of bodies and broken things."

Lee nodded and said, "So, Naval Intel has to assume it's being done on behalf of the rebels."

"They haven't really made that assumption," said Danner. "At least not yet. If they were sure of that, they would have sent a battle group in there and seized the operation. There's no separatist presence in the system, and so nothing to stop the Alliance from reasserting control."

"So why haven't they gone in yet?" asked Lee.

"It's because of the uncertainty," said Danner. "If it turns out to be a problem, the Alliance can always go in with force. In the meantime, the powers-that-be want to know what he's up to. And that's where you two come in."

CHAPTER 57

DEFINING THE MISSION

"Finally," said Lee.

"Your mission", said Danner, "is to go undercover, infiltrate Chambers' organization, make contact, find out what he's doing, and report that to me."

"And then what?" asked Lee sharply. "Clearly, just knowing what he's up to isn't the end game of Alliance Naval Intelligence."

"Fair enough," said Danner. "Intelligence is never the *end game*, as you say. It's always a tool for operational decision-makers."

Lee was not happy, and it showed on her face and the tone of her voice. "Why this team? Why you, me and Laura?"

"You are his niece, his former aide de camp and his chief of staff," said Danner. "You can make a connection where others might not be able to."

Lee shook her head, clearly unsatisfied with that answer.

"Laura is here because she is Nemeth's wife," said Danner. "It was thought that her presence would facilitate any contact with Nemeth."

"And you?" asked Lee. "Why are you the handler? You're right out of the academy and, as far as I know, you have zero experience in undercover operations."

Danner nodded and said, "You are completely correct on that point. This entire operation is a compromise."

"You'd better tell us then," said Lee.

"I am trained as a strategic intelligence officer," said Danner. "I have no training and no experience in covert ops. A typical handler would have extensive training and would be a veteran of years of undercover operations prior to becoming a handler."

"Why you then?" asked Lee.

"When it became known that Chambers had done what he had — apparently gone rogue and set up a criminal organization — I was tasked to be a member of what is called an ad hoc options group. As you can imagine, I was the junior member by far in the five-person group. The idea was to develop options and present those options to the head of Naval Intelligence for her approval."

Lee rolled her eyes, and said with sarcasm, "Let me guess; an idiot *committee* decided this course of action."

Danner smiled wryly, shrugged and said, "Not exactly. The *idiot committee* presented options to the chief of Naval Intel. The decision was made by the chief and, I assure you, she is not an idiot. Admiral Bluefield is as competent as anyone."

"So, what were the recommendations?" asked Lee.

"First, the committee suspected that one of the other intelligence agencies was running Chambers," said Danner. "As you know, the Alliance intel communities do not cooperate so much as compete. And since the Attack of December 7th, almost all sharing has shut down. It was believed entirely possible, even likely, that Chambers was already an Alliance asset. Discrete inquiries were made, and nothing came back. It wasn't ruled out, it just seemed unlikely based on the responses we got back from trusted sources in other agencies."

Danner paused and looked to see the other two were tracking with his explanation.

"The majority of the committee, in fact the other four, wanted to go in with a battle group, occupy the system, seize whatever organization Chambers was running, and arrest Chambers and Nemeth."

"Arrest them?" asked Zakany sharply.

"Even if they are not doing anything illegal," said Danner. "They are, in fact, absent without leave, and have been for over two years."

"So, it was four against one," said Lee. "What was your recommendation?"

"I argued that Chambers and Nemeth could not be doing this for profit. Just no way. Not only was that contrary to their values, but it also didn't make any sense. There are far easier ways to make money, legally or illegally, now that the economies of the Alliance are in chaos. Just look at John Raymond. He started a legitimate research company and is now worth billions."

"And did this convince anyone?" asked Lee.

"No, it did not," answered Danner. "The members of this committee have seen so many betrayals, so many senior officers turn bad, they didn't buy the assertions of a young intelligence officer, who they felt was perhaps star struck by a charismatic leader."

"Again, what was your recommendation?" asked Lee.

"I wanted a full up undercover operation, fully supported and run by experienced personnel. I didn't expect to be involved, nor did it ever cross my mind that you two would play any part of that operation."

"So did your recommendation survive the committee's report?"

"It was four to one. I insisted my recommendation be included as a dissenting view," said Danner.

"And with what result?" asked Lee.

"After getting the report, Admiral Bluefield called me in alone," said Danner. "She heard me out, let me argue the point that Chambers had to have a valid reason for taking over that operation and that we should tread carefully, so as not to disrupt whatever he was trying to achieve."

"What did she say?" asked Lee.

"She shocked me," said Danner. "She said she already knew why Chambers was in that area. When I asked her how she knew that, she told me that signals intelligence had intercepted and monitored a video call between Chambers and John Raymond, now the head of one of the largest and most powerful think tanks and defense contractors in the Alliance."

"Holy crap," said Lee. "Eavesdropping like that has got to be illegal."

Danner nodded. "When I asked her why that information hadn't been shared with the committee, she just smiled as if to say, '*You know the reason.*'

"She said, 'Lieutenant Danner, neither of these recommendations is acceptable. We can't go barging in with main force without blowing any advantage we might have, especially since we can't be sure what Chambers is up to.'"

"And your recommendation?" asked Lee.

"She said that she couldn't do a full-up covert operation because the number of people involved would make a leak almost certain. Sorry to say, that's the world we live in. Everybody is watching everyone else. Even the redeployment of personnel assets that would be required for a full-up covert operation would trigger unwanted attention."

"So, the solution was the three of us?" asked Lee in astonishment. "Three amateurs with no support?"

"Yes, exactly," said Danner. "We're it."

CHAPTER 58

A SONIC BOOM

As Lee walked the paths in the woods, the woods she now thought of as hers, she was frustrated. She had been pulled out of company command for this mission, an undercover operation with no real guidance or support, no intelligence, no extraction team if anything went wrong, and no training.

What a fool's errand. And what made it worse was that General Motubu had agreed to release her to the Navy for this absurd mission. Had he known what a pile of crap this mission was?

It also bothered her that she couldn't even get the mission going, because they were stuck on this planet, a planet that didn't have electricity, much less interstellar communication.

Danner had told her that a distress message should have gone out when the shuttle began to break up. Ali should have sent one, and if not him, the ship's AI would have done it automatically.

But they had been here for three months, and there had been no sign of any rescue. Danner said if the distress call did not go out, or had not been picked up, the next trigger would have been his failure to report back to Admiral Bluefield. But he was not expected to do that until about

90 days into the mission. That meant that Admiral Bluefield would just now start wondering where the stray Lieutenant Danner had gone missing.

What infuriated Lee was that Danner had no real plan for inserting her into her undercover role. He said they would figure it out once they got into the system.

Lee was used to meticulous planning, with scrutinized courses of action, contingencies, and abort criteria if things went wrong. This whole thing was just a mess.

What made it worse was that the Marine Corps actually had special operations units designed for deep cover. Lee had glimpsed them once when she was in training. She had seen in one of the adjacent training areas, a group of men and women in civilian hiking clothes. The men had long hair and beards, the women also with hair, dress and demeanor well outside of standard Marine Corps protocol. Lee had asked the instructor why civilians were in the training area.

He said, "They're not civilians, they're Marines getting ready for some type of black ops." That's all the instructor would say, and Lee didn't press the issue further.

But later she asked around and heard in whispers that these Marines were part of a counter-terrorism unit called Oracle, that could go undercover for their missions, pretending to be civilians.

Lee remembered that Admiral Chambers had served in Oracle along with Nemeth back in the day, before Chambers had been injured and forced to switch to the logistics corps. Oracle, she recalled, was a force made up of Naval and Marine special operations personnel.

I bet if they had been given this mission, Oracle wouldn't be stuck on this planet for months waiting for a rescue mission, thought Lee.

Just then she heard the unmistakable sound of three sonic booms in quick succession. She looked up and saw the streaks of starships entering the atmosphere. She started running for the monastery.

As Lee entered the village at a run, she could see the Mongol guard was on full alert. Archers were on the wall, and troops were moving to their rally points strapping on armor and checking weapons.

As she passed the gate for the innermost wall she was met by Sarnai, Aruban, and a squad of his troops. Lee opened her mouth to speak when a roar sounded directly overhead. She looked up and saw a shuttle flying about 50 meters above the village. She only got a glimpse. It was a shuttle, but it clearly wasn't a Naval spacecraft. *So, not a rescue then*, she thought.

She turned to Aruban and said simply, "These aren't my people."

"Understand," he said. "I assume if they can travel between the stars, they have powerful weapons, yes?"

"Yes," she said. "You won't be able to stand against them if they are hostile."

Aruban said in a steady voice, "The guard will remain and fight them if necessary. The civilians will evacuate."

"Evacuate?" said Lee. "Where will they go?"

"The Olgoi have dug us tunnels for just such a time as this. The people will disperse to the hills."

"You planned for this?" asked Lee in surprise.

"Yes, of course," he said. "Once you came, we knew it was possible for others to follow. The Olgoi have been good allies."

As they watched, the three craft set down, side by side, about 100 meters in front of the main gates of the village.

"Let's go meet them," said Aruban.

CHAPTER 59

A CONFRONTATION

Lee stood next to Aruban and Sarnai, with a squad of guards behind them. Lee was in full battle armor, carbine in hand, pointed downward. They waited about 30 meters in front of the three shuttles. Mongol archers lined the village defensive wall behind them.

Now that she was closer, Lee could see the three shuttles more clearly. As she suspected, they were not Alliance military spacecraft. These were civilian craft, but they had been subtly modified with additional armor and weaponry in such a manner as to hide the modifications from anyone making a casual observation. Lee could spot the changes only because of her years of experience as a Naval officer. One of the three was larger, and it had some type of external equipment that Lee could not identify. That was the one that had buzzed the village.

She knew that the Mongols were completely outgunned. She could see that the three shuttles mounted recessed 30-millimeter cannon, which, if fired, would obliterate her, the Mongols and the defensive wall. She took it as a good sign that the craft hadn't opened fire yet, something they surely could have done had their owners been so inclined.

She looked to her side at Aruban and saw a man who was completely calm in the face of potential death. For him, duty was all. Facing these strange spacecrafts was simply the correct thing to do. The consequences to his person were secondary.

Lee also knew that Aruban was stalling for time. The evacuation was now ongoing, but the tunnels could only take so many people at any one time. His job was to protect the people of the village. By confronting the three strange craft, he hoped any aggressive action on the part of these visitors would be delayed long enough for his people to get away.

Lee had asked Sarnai, "Will the monastery evacuate?"

Sarnai shook her head and said, "Some of the civilian support staff will evacuate. The monks and sisters will not. The monastery is sacred ground, a sanctuary, and they will not leave it."

The ramp lowered down from the middle of the three shuttles. Out walked a man. At first, Lee saw only his legs, long and strong, then the torso, then the shoulders and head.

The man was tall, dressed in simple black jeans and denim jacket. He had a short, trimmed beard with longish hair falling to his collar. He was slim, but powerfully built. He walked with confidence and poise, not arrogance. Lee knew the walk. She had seen it before.

He walked directly to Lee, stopped three paces in front of her, came to attention, saluted, and said, "Ma'am, Lieutenant Theodorus Wasp reports."

Lee was stunned, she opened her mouth to say something, closed it, then realized Wasp was holding his salute waiting for her to return it. She snapped a return salute.

Here was the man, the only man, she had ever let into her life. For a brief time, she thought he was the one. They had murmured words of commitment. And then he was gone. Not a single word from him in over two years.

She paused a long moment and then said, "It took you a while. We've been here for three months."

Wasp said only, "Yes, ma'am."

Lee turned to indicate her companions. "Lieutenant Wasp, Alliance Marine Corps, may I introduce Commandant Aruban, head of the local constabulary." Aruban nodded to Wasp.

"And Prioress Sarnai of the Ongii Monastery." Sarnai gave a slight nod, and just a hint of a smile at the tall Marine.

"Sir, ma'am, it's my pleasure to meet you," said Wasp as he nodded.

Lee looked past Wasp where there was movement at the shuttles and said, "Who are your friends?"

He turned back toward the shuttles, and they watched other people coming off the spacecraft. Some hung back, but two came forward: a middle-aged man and a wiry, tough looking woman of about 25. She had a shaved head and a tattoo of a dragon on her right forearm. Lee recognized them both.

As the two joined Wasp, the woman came to attention and saluted, saying, "Ma'am, it's good to see you again."

"Private Johnson, it's good to see you," said Lee as she returned the salute.

The woman smiled broadly and said, "It's *Sergeant* Johnson now."

"Congratulations on your several promotions," said Lee.

Lee remembered Jessie Johnson for two reasons. The first time they had met, Lee was a Naval officer who had been invited to teach a martial arts class to a group of elite Marine Commandos serving on the Achilles Battle Fleet. Johnson was a judo expert who, at Lee's request, had demonstrated a technique that threw Lee head over heels and onto her back.

The second time was during the insertion into Alpha 51. The task force Lee had been commanding was to launch individually from a spinning meteor. Johnson's propulsion device had failed, and she had to remain on the asteroid for hours before being rescued. By the time the Marine Search and Rescue team had reached her, she was unconscious and barely hanging on to what was left of the disintegrating asteroid.

Once revived, she had refused further medical treatment, and insisted on joining the S and R team. She helped dig Wasp and the rest of her team out of a collapsed building on the moon where they had been trapped.

On both occasions Lee had been impressed with her toughness and positive mental attitude. Women Marines were rare, and rarer still in the elite commando units.

The man was also someone she knew from her time in the Navy: Chief Warrant Officer John Raymond, now retired.

"Chief," said Lee. "You are absolutely the last person I expected to see coming out of that shuttle. It is good to see you."

Lee turned back to Wasp. "We have a lot to discuss. But first, are you here to rescue me or arrest me?"

Wasp laughed easily and broke into that smile that she remembered so well. "Neither," he said. "We're here for a mission, *your* mission."

"Glad to hear it," said Lee. "But I need something from you right now."

"Tell me," said Wasp, his face now serious.

"You have a medical team and an autodoc, I assume?" asked Lee.

"Yes, we do," said Wasp.

I have two Naval personnel in comas, and two others with residual injuries. "We need your help now."

CHAPTER 60

MODERN MEDICINE

Lee was impressed with the medical team that came with Wasp. It was comprised of three Navy corpsmen, all petty officers. They brought their equipment in specially designed backpacks and moved quickly through the village and into the monastery's infirmary.

They were met by Dr. Zakany, who explained their symptoms and the treatment they had been given. They quickly set up their equipment, started an IV and began tests on the two coma patients.

Part of their equipment included a monitor with a visual display. Some of information displayed, Lee had seen before and recognized: a heart monitor, blood pressure, pulse, an EKG and an EEG.

"Blood work coming through now," said one of the corpsmen.

"I see it," said Zakany. "Neisseria Meningitidis, or something close. Let's see the scans."

On the display, Lee saw the scans of the various structures of the body: internal organs, brain, muscles and skeleton. Lee didn't understand any of it. Zakany and the two medical technicians looked that the screen, pointed to certain parts of the display and spoke in low tones, occasionally changing views to different images and read outs.

Lee heard Zakany say, "We are agreed then?" Both of the corpsmen nodded, and one responded, "Yes, doctor."

Zakany turned toward Lee and Danner. "Both Ali and Odessa have a form of bacterial meningitis. This is almost certainly from the bites."

"What's the treatment," asked Lee.

"Massive doses of antibiotics and antivirals," said Zakany. "We'll see how they react to those. In about a week we'll know more."

"Will they recover?" asked Danner.

"I don't know," said Zakany. "This is a new form of the disease, it's one of the reasons it hit them, and you, so strongly. The body has no natural resistance. You have all been inoculated against meningitis, but not against this strain of it."

"How long before they can be normal again," asked Danner.

"There has been some damage to the internal organs, the central nervous system, and the brain. If we can clear out the bacteria, the source of the disease, we can treat them and try to repair the damage. Unfortunately, that cannot be done here with what we have. We will have to stabilize them, and then get them back to a suitable medical facility."

Zakany looked at Danner, "As for you, Lieutenant Danner, we'll need to run a full check on you as well. Somehow, even though you were bitten, you recovered more quickly. I can run tests on you tomorrow if you will come by in the morning."

Danner nodded and said, "These two worked for me. I am responsible. Is there anything I can do to help?"

"Yes," said Zakany. "Continue to heal."

The next day, Danner underwent a full physical exam. It took two hours. In addition to the technical aspects, Zakany had him perform a number of manual tasks. He walked a straight line, closed his eyes and touched his nose, stood up from a sitting position, followed her finger with his eyes without moving his head, counted backward from 100 by sevens, and identified images presented to him – elephants, chairs, shoes, a baby.

After it was over, he was clearly exhausted.

"So how did I do, Doc," he said with a weak smile.

Zakany was not smiling. "You have moderate damage to your central nervous system," she said simply.

"How bad?" he asked.

She touched a screen, and an image popped up on the monitor. It showed an image of a spine with the many branching tendrils. She pointed to a place on it and said, "This here is a lesion on the spinal column. It's basically a scar from the meningitis. You have these throughout your body, but this is the worst of it."

Danner showed no emotion at this and simply said, "What's the implication?"

"Clearly you are weak from fighting the disease, your immune system is struggling, and your body is trying to repair the damage," said Zakany. "Although, you haven't shown it, I can see from the results you must be in substantial pain."

Danner shrugged.

"Remember, Danner, I'm your doctor. Fess up," she said with a smile.

"I am a Naval officer," he said. "I'm supposed to be in charge of this mission. I can't be whining about aches and pains when people I lead are much worse off."

"I understand, and that is commendable," said Zakany. "There is the issue of your cognitive abilities."

"I thought I was doing all right in that area," said Danner.

"It's a mix of things," said Zakany. "The good news is your cognitive ability is still higher than average, even for a Naval intelligence officer." She smiled and then continued. "But you are way off your normal self. You have an engineering degree and a degree in ancient history, all earned with honors by the time you were 21-years old. When you solved a problem on the Achilles Battle Fleet by sifting through sensor data to isolate quantum effects, Chief Raymond said you ought to get a Nobel prize. He has two himself, and he wasn't kidding."

"Your point?" asked Danner.

"My point is that you just had trouble counting backward from one hundred by sevens. That failure wouldn't be a problem for most of us, I probably couldn't do it without a mistake. But the Andy Danner before this recent trauma, could do multivariate calculus in his head."

"If I'm still smarter than the average bear, why is this a problem?" asked Danner.

"It's a problem because you rely on your superior intelligence in your everyday life," responded Zakany. "You don't have the coping skills, memory aides, and shortcuts the rest of us have that we use to deal with our less-than-perfect memories."

"So, what are you saying, doctor?" said Danner with an expression of anger on his face.

"Let me ask you this," said Zakany. "During the snatch operation with the Achilles Battle Fleet, you stayed at your post directing a rapidly developing and unique combat operation, while the command center was breached, and half your team were killed around you. You didn't make a mistake and you didn't flinch."

"Your question?" asked Danner.

"My question is – could you do that now?"

Danner opened his mouth as if to answer and then stopped, closed his mouth and sat in silent contemplation.

Finally, he said, "I'm no different than any other sailor or Marine who has been wounded in combat. This is my mission, and I intend to complete it as long as I am able. I hope you're not intending to waive your magic doctor wand and take me off this mission."

Zakany looked at him with doubt in her eyes.

Danner said, "When Mei Ling was wounded, seriously wounded, during the battle of Alpha 51, she lost consciousness twice from concussion, she had broken ribs. *You* were the medical authority, and *you* cleared her to continue with the mission."

Laura blushed at the memory, nodded and said, "You're right, Andy. I did that. But do recall that she voluntarily took herself out of the leadership role because she knew her judgment was impaired. Think about that."

"Are you going to take me off the mission for medical reasons?"

"No," she said. "I'll leave that up to you. But I will make a full report to the others on the medical condition of all of us. If the issue is raised with me, I won't lie."

"And the others?" asked Danner.

"Ali and Odessa need to be evacuated to a proper medical center. You don't need to be evacuated. But you won't heal completely with the resources I have here. I'll put you on antibiotics and antivirals, but the damage to your spine requires a neurologist and extended therapy to heal. The longer you wait, the greater the risk that damage will become permanent."

"Thank you," said Danner. "Thank you for your care of me, and of the others. Going into combat, I couldn't ask for better."

Surprised at the new tone, Laura Zakany blushed furiously and said nothing.

CHAPTER 61

A CHANGE OF COMMAND

Lee, Zakany, Danner, Wasp, Johnson and Raymond sat around a large round table in one of the monastery's rooms. A fire blazed in the hearth.

Lee had requested from Sarnai a place for this meeting. Sarnai had seemed to understand it was an important meeting, and she had assured Lee that her privacy, and that of her guests, would be respected.

Wasp had a laptop in front of him, as did Raymond. Lee, Zakany and Danner had nothing; they had had nothing electronic for months since their crash on this planet. Lee was pondering how they had adjusted to not having electricity at all: no lights, no phones, no computers. Her suit had a residual power supply, but she had no use for it since the battle against the Nephilim. *It wasn't that bad,* she thought. *We didn't really miss it.*

Since Danner was in charge of the mission, Lee waited for him to start.

"Thank you for coming to our rescue," said Danner indicating Wasp, Johnson and Raymond.

"Before we start, however, I need to make something clear. I was tasked by the head of Naval Intelligence, Admiral Bluefield, to carry out the mission and all supporting tasks. I am obliged to do my best in this assignment, and to answer back to her on the mission accomplishment."

He looked around again, as if waiting for a challenge.

"That said," continued Danner. "Doctor Zakany has informed me, and convinced me with sound evidence, that my bout with this alien form of meningitis has taken a toll on my physical and cognitive abilities. Should we face stressful combat situations, my judgment could be impaired. And I will not allow such a situation to occur. Therefore, I am turning over command to Captain Lee, as she is the senior officer present. I will support her and this mission to the best extent of my abilities."

He looked around again and saw the understanding and respect of those present. Nothing is harder for a leader than to hand over command to another.

He continued, "Captain Lee will determine my participation in this mission; I will take my orders from her. My only caveat is that I will not allow myself to be medically evacuated with the pilots. I may be impaired, but I am stable, and I *can and will* be useful. I will stay on this mission."

He turned to Lee and said, "Captain, the meeting is yours."

CHAPTER 62

MISSION RESET

Lee paused and looked around the table, taking a moment to look each of those present in the eye.

"Let's start with the original mission parameters," said Lee. "Then let's go around the room and hear what everyone has to add."

When no one objected, Lee said, "I received a very brief and somewhat cryptic task from General Motubu. He pulled me out of company command and told me only that the mission was important and that a Naval Intelligence officer would arrive soon with the information on the task. That officer was Lieutenant Danner, who had with him Doctor Zakany. We departed in his shuttle, leaving his security team behind because, he told me, they were not cleared to know about or participate in the mission."

She paused again and then continued. "Before he could tell me about the mission, we were attacked by pirates. In dealing with them and in making our escape, our ship took damage that caused us to enter this planet's atmosphere in distress and then to eject. I was recovered by the monks of this monastery, while Laura, Danner, Ali and Odessa were captured by a species the locals call the Nephilim."

"The Nephilim laid siege to the city and brought their four captives with them. I led a party of Mongol soldiers to attempt a rescue. During that effort, another species, the Olgoi, who had been captives of the Nephilim, revolted and came to our aid. As a result, the four Alliance personnel were safely recovered, and the Nephilim were killed or driven away."

Lee paused and then continued, "Since his recovery, Lieutenant Danner has informed me that the mission was for me to go undercover to find out what Admiral Chambers is doing on Ulysses in the TOI-700 system."

Lee turned to Wasp and said, "Lieutenant, can you tell us your story?"

"Ma'am," he said. "Yes, I can. I don't need to remind all of you that what I am telling you is classified at the highest level, nothing I say about my mission or my unit can *ever* be revealed to others, no matter their security clearance."

He looked around and saw that there was agreement from all. "I am currently the team leader for Oracle Group Alpha. We are black ops. Neither the budget, nor the personnel, nor the mission is releasable outside a very small and secure group. Sergeant Johnson is my deputy. We have a total of 12 marines and Naval personnel on this mission. We have three shuttles, all of which are civilian craft upgraded to our specifications. Both the craft and the personnel have verifiable civilian identities, should we ever have to interact with civil or military authorities."

He paused and looked at Lee. "Questions?"

"How did you get on this mission, and what are your orders?" asked Lee.

"I was ordered by my group commander to report urgently to the Commandant of the Marine Corps, General Motubu. I did so. He told me that you, Captain Lee, had been released by him for a classified mission. He had agreed to that based on a request from Admiral Bluefield, Chief of

Naval Intelligence. At the time he agreed to that transfer, he did not have a full picture of the mission, its requirements, or of the constraints."

He paused, and Lee leaned close. She wanted to know why Motubu, whom she trusted, had agreed to this shit show.

"When it was clear that your shuttle had gone missing," continued Wasp. "General Motubu reached out and made discreet inquires as to the mission and what efforts were being made to rescue or recover your team. What he found out was that no efforts had been made to find you, and that the original mission had been allocated zero support from the Navy."

"And that's when he reached out to your group," said Lee. It was a statement not a question.

"General Motubu called me in," said Wasp. "He explained about your mission and that he had seconded you to the Navy for an undercover mission. He said you had gone missing and he wanted my team to find you if we could. But he also wanted us to engage the mission to contact Admiral Chambers.

'So where have you been?" asked Lee with an edge to her voice. "We've been *missing* for over three months."

"It took a while to track down what had happened. We found the pirates who boarded and tried to take your shuttle." Wasp smiled and said, "They were pretty forthcoming when I told them you expected their full cooperation. Apparently, you made an impression."

Lee shrugged her shoulders and said with a tight smile, "They were very rude."

"Undoubtedly," said Wasp. "So, we knew you had gone into warp drive partially damaged. As you know that is universally considered a fatal condition. No known vessel has ever been recovered under those circumstances."

"But you didn't give up," said Lee.

"No, we didn't," responded Wasp. "We used your final trajectory and plotted every possible location where you could have come out of warp. There were 53 locations in total. About half of those were not survivable. We eliminated those from our search criteria. This planet is the seventh one we've checked,"

"How did you know we were here?" asked Lee.

"We couldn't do a proper scan; there's something in the planet's crust that causes electromagnetic interference. But luckily there is still some of the debris from your shuttle in orbit," said Wasp. "And your battle armor still has a residual charge that we could track when we got closer."

"Is this your entire team?" asked Lee.

"Not all," said Wasp. "As I said, we were tasked with completing your mission, so we have taken steps in that direction in case we couldn't locate you or if you had been killed."

"Steps?" asked Lee.

"We have inserted an advance team of two persons a month ago," said Wasp.

Of course, thought Lee. *Oracle has its act together. They would have people trained and ready to go undercover.* She was furious that she had been sucked into such an amateur operation. It looked like Wasp had saved the day once again.

Lee turned to Raymond. "Well, *Mr.* Raymond. How does a civilian come to be along on a covert special ops mission?

CHAPTER 63

RAYMOND TALKS

"Actually, General Motubu came to see me at my headquarters on Maui," said Raymond. "He just showed up unannounced in civilian clothes, and asked at the front desk if he could meet with me. Because our business is mostly defense contracts, our people keep a database of senior military officers. When he showed his ID at the desk, I was immediately notified and came down to greet him."

"That's interesting," said Lee. "Was he trying to keep a low profile?"

"Yes, he said he was actually on vacation, and that he liked visiting the Hawaiian Islands. But he is a Marine, so he quickly got down to business. He told me his intel service had monitored the call with Admiral Chambers. He explained that Lieutenant Danner, Captain Lee and two crew members had gone missing in an effort to make contact. He said he had tasked a Marine special ops team to try and find them, but that he hoped I could be of help since I was the one to point Admiral Chambers in the correct direction."

"What did you say?" asked Lee.

"I told him what I knew. He seemed happy with that, but that's when I said I wanted in on the mission."

"What did he say to that?" asked Lee.

"At first, he said, *no way*, that I was a civilian. I reminded him that as a defense contractor, I had a security clearance as high as almost anyone still in the military. I also made the argument that I could be helpful once we contacted Admiral Chambers. And finally, that I had the resources that could help disguise the movement of his spec ops team."

"What resources?" asked Lee.

"This entire operation is being funded by RRA," said Raymond. "Those shuttles are mine, all modified to spec ops requirements. They don't have to pretend to be civilian craft; they really are."

Wasp added, "Chief's support helped us out as well. One of the reasons Admiral Bluefield didn't want a full up support team for this undercover operation was that the movement of so many resources would alert others in the intel community, and not all the actors in this field are trustworthy."

"How did his support of the mission make a difference?" asked Lee.

"Nothing is coming out of any military budget. It's not just off books, there are no books. The military team with me, all of us are on annual leave."

"I'm impressed," said Lee. She looked at Raymond closely. He had changed quite a bit since she had first known him as a chief petty officer who worked as the maintenance chief for the 514th recon squadron. At the time, he had seemed to fit the type: gruff, competent, maybe a little heavy. She would *never* have guessed he was the missing scientist, Dr. John Raymond.

Raymond had disappeared from public view some 20 years prior. Most people believed that he had either died or that he had gone into some type of self-imposed exile. No one had imagined that he had joined the Navy as an enlisted sailor to avoid government assassins. At the time of his disappearance from academic life, he had been part of a classified research project investigating a phenomenon called macro-wave collapse.

After the project had been forcibly shut down by the government, many of Raymond's former colleagues from the project had suffered fatal accidents or gone missing. Raymond, believing his life was in danger, simply joined the Alliance Navy, changing his middle name, and then spent the better part of the next 20 years in space.

When the war broke out, and the mysterious enemy had begun using a technology that Raymond recognized as being a derivative of his macro-wave collapse research, he revealed himself to the Admiral, and he set about trying to discover ways to protect the fleet from attack.

An internal spy had attacked Raymond and severely injured him, after killing two of the Marines tasked with guarding him. When the fleet had returned to Earth, Raymond had retired from the navy and set up Raymond Research Associates or RRA, which quickly became one of the most powerful and influential defense contractors. Raymond was now a very powerful and wealthy man. *What was he doing here with a black ops team?* Wondered Lee.

"So let me start by saying I had contact with Admiral Chambers and Warrant Office Nemeth within a few months of their disappearance following the battle of Alpha 51."

He paused and breathed out as if gathering himself. "I am not betraying a confidence by saying this, because I now know that several intelligence services monitored that call, including both the Navy and the Marine Corps."

He shrugged and said, "Clearly, I need to up my game as far as communication security. The admiral had asked me to use my company and my own analytical skills to find where the enemy would make strategic efforts now that the plot was exposed with the capture of the quantum flux generator on Alpha 51."

"Wait a minute," said Danner. "Why all the subterfuge? Why fake his own death?"

"Because he understood that the fight against those who attacked the Alliance from within was ongoing," said Raymond. "That the powers that be would try to stop Chambers, as they tried to stop Captain Lee when she returned."

"Don't get me wrong," said Danner. "I respect Admiral Chambers. He was a great leader at a very difficult time. But I have to say, it looks like he left us exposed after the battle. Let's not forget, Mei Ling was almost arrested, and she did have to undergo a court martial."

"All true," said Raymond. "But he did reach out to both the Chinese Consulate and to General Motubu to make sure Mei Ling would not be facing these problems alone. And there's no guarantee his presence would have made it any better for any of us."

"So, what did your research turn up? What is so interesting about TOI-700?" asked Lee.

"We found that a criminal organization had taken over the governance of Ephesus, and had begun mining operations in the system's Oort Cloud."

"I understand they are mining some type of rare materials?" asked Lee. "Why is that important?"

"Boron Nitride and other related materials are extremely rare and are used for quantum computing," said Raymond. "We knew they would need this material for the computations required for macro-wave collapse. But when I spoke to the admiral, we still didn't understand why they would need so much. Projecting material objects from one place in space to another is limited by the lack of information inherent in great distances. More quantum computing doesn't solve the problem."

"So why did he go there?" asked Lee.

"I think he believed it was just the best place to start, and by going there he would be in a better position to stop or interdict whatever they might be planning," said Raymond.

"So, what have you found out since he left for this mission?" asked Lee.

"We've poured a ton of resources into the issue. We also have our own commercial intelligence assets, all run by former Alliance intelligence officers and NCO's," said Raymond.

Lee inwardly sighed as she felt a lecture coming on. But there was nothing for it but to let Raymond do his professorial thing.

Raymond then surprised them all by saying, "Time travel."

CHAPTER 64

TIME TRAVEL

Silence met this assertion. Raymond looked around as if waiting for someone to challenge him.

Finally, Lee said, "Tell us, Chief."

"First, some background," said Raymond.

Lee could feel the others squirm in discomfort at the lecture they knew was coming.

"One of the tech companies I acquired when I was building RRA is a small firm which had been dedicated to pure research into a single topic, and that was parallel universes."

The people in the room stared at this, either excited or uncomfortable with the topic.

"As you will be aware, one of the paradoxes of quantum physics is that when a particle is in its wave form, it is in a superposition state that allows it to be everywhere at the same time. When the particle is detected, or it interacts with some other entity, then the wave function is said to collapse. The particle is no longer in a superposition state; it is in one place only."

Raymond looked around to see if everyone was following. There were some nods and some blank stares. Raymond decided it was enough to continue.

"The multiverse theory states that every time the wave function collapses, all of the infinite possibilities of where the particle could be are, in fact, realized in an infinite number of universes that are *created,* by the choice made by the detection of the particle in a specific place and time. We see it as a collapsed wave function, but that is only because we go with the reality that was created with that choice. The multiverse theory says that there are many versions of us and our world, that split off from us when the decision is made."

"What does this have to do with time travel?" asked Danner.

"Getting there," said Raymond a bit sharply. "The appeal of the multiverse theory is that, if it were true, it would solve the apparent paradox of superposition. Without going into too much detail, superposition violates both special relativity and local causality."

"What's the problem, then?", asked Lee.

"The problem," said Raymond, "is that until now, there was zero physical evidence of the parallel universes.

"And that has changed?" asked Danner.

"Yes," replied Raymond. "One of the reasons I took note of this research firm in the first place is that they had been buying an inordinate quantity of Boron Nitride. At first, I suspected they might have something to do with the TOI-700 mining, but that turned out not to be the case. Their purchases turned out to be completely legitimate on the open market."

"So, what were they doing?" Asked Lee.

"They had made a breakthrough, and they were begging for investment money to keep going," said Raymond. "The few scientists working on the project had mortgaged their homes and borrowed from relatives to keep their research going. When I found them, they were sleeping at their lab and living off peanut butter and jelly sandwiches."

"So, you bought them out?" asked Lee.

"I acquired their firm as a subsidiary, yes," said Raymond. "The scientists all stayed on, they were well compensated, and they have an equitable interest in any patents or royalties, should those ever materialize. But most importantly, I funded their research and the results have been substantial."

"They discovered time travel?" asked Danner.

"Yes and no," said Raymond. "What they did discover was a way to see into parallel universes, those that are very close to ours."

"How is that possible?" asked Danner. "I thought the parallel universes were completely separate, that no information could ever be shared between any of them."

"Yes, that's what we all thought," said Raymond. "But these guys figured that these alternate universes couldn't be completely separate because at one point they were together as the same universe. The key was to find the exact point of separation, and to follow the alternate universe from that point of fracture."

"But I thought these worlds didn't interact with each other," said Danner. "How did they find this point of separation?"

"That was their genius," replied Raymond, who was starting to show his enthusiasm. "What they figured is that the parallel universes do interact with each other."

"OK. Please explain," said Danner.

"They looked for any situation in which a single particle shows evidence of being interfered with when there are no other particles near it," said Raymond.

"You're talking about the two-slit experiment," said Danner.

"Exactly," said Raymond. He looked around and saw that the others at the table didn't follow the argument. He nodded and dove into an explanation.

"We've known for centuries that if you fire photons, or even particles having mass at a barrier that has two slits, the resulting pattern on a subsequent detection plate will show a wave-like pattern, rather than a shot group one would expect if the particles or photons went through one slit or the other. This is called an interference pattern. The standard interpretation of this is that the particles pass through the slits as waves and interfere with each other."

"Got it," said Danner.

"The problem arose when the experiment was limited to a single photon being fired at the two slits. One photon or particle at a time. When that happens, the resulting pattern on the detection plate is still a wave pattern. It was as if the single particle interfered with itself."

"So? How does that help find the split?" asked Danner.

"Our guys reasoned that the single particle was being interfered with by other single particles from a parallel universe. This is the point where the two or more universes interact, and that's where they focused their efforts."

"So, they found it, another universe?" asked Danner.

"What they found is that these alternate universes aren't really separate. They are here, with us."

"Why can't we see them?" asked Danner.

"It's because the space they exist in is *folded* in a way that we can't see it," said Raymond. "What these scientists did was to use their massive supercomputers to *find* the opening of the fold and then follow it through to the other universe."

Danner said, "But how does that get us to time travel?"

"The key is that these parallel universes are infinite, and some of the timelines are offset with each other," said Raymond.

"What does that mean?" asked Danner.

"It means that once we can peer into the multiverse, we always find at least one parallel world that is nearly identical to ours, but at a different point on our timeline."

"So, you're saying we can look into a parallel universe that is just like ours and see what happened on a different date?" asked Danner.

"Not just see," said Raymond. "We can go there; we can travel to that time in a universe that is so nearly like ours that the differences are mostly at the subatomic level."

CHAPTER 65

TIME TRAVEL, BUT NOT REALLY

"Got it so far," said Lee. "But how does that get us to time travel? Moving into an alternate universe at a different point on the timeline doesn't send us back in time in our universe. I don't see time travel."

Raymond nodded as if Lee were an exceptionally bright student. "You are correct. Time travel in one's own universe is impossible. It would create what is called the grandfather paradox."

Raymond could see the others didn't understand the reference. "The grandfather paradox occurs when a person goes back in time and kills his own grandfather while the grandfather is still a child. If the grandfather is killed before he can have children, the time traveler is never born and cannot got back in time to kill his own grandfather. Thus, the paradox."

"So, no time travel," said Lee with apparent frustration. "I don't get it. You said your people had discovered time travel."

"That's why I said *yes* and *no*," said Raymond. "But the effect is the same. Remember, these universes are nearly identical. That means there are alternate versions of each of us."

Danner opened his mouth with an expression of astonishment on his face, "You mean..."

"Yes," said Raymond. "Whatever we do here, other versions of us are doing in their universe."

"So, when we travel to an alternate universe and insert into another timeline," said Danner excitedly, "Another version of us is entering our timeline at the same point."

"Yes, exactly," said Raymond.

"You have confirmed this?" asked Danner.

"Yes, over a dozen times," said Raymond. "The easiest way is to go back in time in an alternate universe, and leave an object buried where it won't be disturbed," said Raymond. "Then we return to our own timeline and dig where we left the object. In every case, it's there. Our doppelgängers have mirrored our actions. When we traveled to an alternate timeline, someone traveled to ours."

"And you've done this?" asked Lee with a look of suspicion on her face.

"Yes we have," answered Raymond. "In fact, I anticipated your lack of belief. You recall my craft buzzed the village before we landed out front?"

Lee frowned. "I was not happy to see that. You caused an evacuation of the village."

"Sorry for that," said Raymond in a way that showed he was not sorry. "We did that to get a precise survey of the village and its walls. I needed to do it by visual flyby because this damn planet has some weird interference from the crust that prevents a good survey from orbit."

Raymond stopped and seemed lost in thought. He said, "Hmmm... yes, that's very odd." Then he looked up and continued, "I then jumped back 1,500 years and planted one of these precisely 12 meters from where the outer wall is now located." He held up a solid block of some sort of shiny metal about two inches square. "Let's go see if it's there."

The group followed Raymond out the gates of the village and past the three walls. It was dusk, and light was beginning to dim with long shadows. Once outside the walls, Raymond opened his handheld device and was clearly navigating by the information displayed. He turned east and walked along the line of the wall, stopped, checked his device, walked a few steps away from the wall and then stopped.

"It's here," he said.

Wasp and Johnson stepped up and began digging with short shovels they unstrapped from their backpacks. In minutes, they all heard the clank of metal on metal. Wasp reached in and pulled out an object encrusted in dirt and rock. He used his Ka-Bar knife to deftly clear away the debris. He handed the object to Raymond, who took it, held it up to the fading sunlight and handed it to Lee.

Lee took the block and turned it over.

"If you test it, you'll find that it is stainless steel, and that it has been in the soil for 1,500 years," said Raymond.

Lee held the block, turning it over thoughtfully. "I don't get it. Where is the cause and effect?" she said. "Did your action of burying it *cause* your doppelgänger to cross over into our timeline and bury this block of metal? What if you hadn't done it?"

"There is no cause and effect,' said Raymond. "The two timelines are so similar, so close in time to the splitting event, that they are virtually identical. The cause for each of the two Raymonds is the same, so the actions are the same."

"Will those actions always be the same, I mean as we move forward in time away from the point of split?" asked Danner.

"It's a good question," said Raymond. "The answer is *no*. At some point, the individual choices of the people involved will start to deviate, and that will have a cumulative effect over time."

"So, was it possible that your double wouldn't have buried this at this point?" asked Danner.

"Not in this instance. When we jump timelines, we can measure how recent the split is. So far, we always picked a time stream that is only a few seconds, maybe a few minutes old. That's close enough to ensure a reciprocal action."

Danner shook his head and put his palms to his temples. He knew something was wrong with the Chief's explanation. The discrepancy was just beyond his reach, but he couldn't grasp it. Zakany was right; he wasn't at his best. He would think on it and hope it would become clear to him later.

Lee looked around and said, "Let's get this mission moving."

CHAPTER 66

MISSION LAUNCH

Lee felt exhilarated. Finally, she was on the move with a mission to accomplish. She was on the bridge of the larger of the three shuttles, which she had claimed as the command vessel. She had made it clear to Raymond that she considered his contribution of the three shuttles as a transfer to the Alliance and, as the senior officer and commander of the mission, she would make all decisions as to their employment. If the shuttles survived the mission, he could have them back.

Raymond accepted her confiscation with good cheer. She had sent one of the vessels back to a special operations base hospital to drop off the two coma patients. In the two weeks it took them to return, Lee took the team through the planning process for the mission. They agreed that the primary objective was to make contact with Chambers to determine his status and intent. If, as they suspected, he was working in the interest of the Alliance, they would seek to assist him. This would need to be done in such a way as to conceal their own identities as Alliance special operations forces, and to maintain Chambers' own cover as gangster.

Wasp had a preliminary report from his two-person advance team: Marines passing themselves of as a married couple. They had processed through the normal immigration procedure, using manufactured identification documents provided by Oracle. For the previous month they had been working in one of Chambers' — Mr. Watson's — semi-legitimate enterprises. It was a sports arena that hosted various types of combat matches, including wrestling, boxing, kickboxing and — for those who were truly blood thirsty — unrestricted fights with no safety equipment and no referees.

These last types of matches were hugely popular, and the patrons paid good money to come watch. They also placed wagers on the outcome. Here is where Wasp had recommended Lee make her entrance into Chambers' criminal world. She could be entered as a competitor in the lower-level matches. By winning, she could work her way up to the unrestricted matches. The idea was that Chambers would take notice of her. If he reached out to her, the first part of the mission would be accomplished.

The team had reported that Chambers had imposed severe controls on all his business enterprises. The worst abuses had been stopped. Prostitution, illegal drug sales, and extortion had all been rooted out, sometimes brutally.

Nemeth, now known as Mr. Attila, was clearly seen as Chambers' deputy. He handled much of the day-to-day operations and the occasional enforcement when subordinates tried to engage in a deviation from Chambers' strict rules.

The team reported that Chambers' cover worked as intended. No one suspected that Mr. Watson was Jay Chambers, former Alliance Rear Admiral. Not that there weren't rumors. Mr. Watson was an enigma to all, and as expected, there were many versions of who he was and where he came from. Some said he was an escaped master criminal, or that he was a former mob enforcer who had struck out on his own. One of the rumors was that he was a cyborg, part human and part machine, as evidenced by his artificial hand.

Chambers himself had encouraged these rumors and had actually started a rumor that he was a former Navy admiral. As he expected, that claim simply joined the many other versions of his background.

But the truth was that nobody really cared. Mr. Watson was the boss, and that was all that really mattered. He was generally well respected because he had established order, and his rules were well known and equitably enforced.

The advance team had not been able to gain any information about the mining segment of the business. And this was the second, and most important part of the operation. Raymond was convinced that mining Boron Nitride was the key to the operation. The advance team knew nothing about any investigations that Chambers might have undertaken.

The plan was for Lee to get into the arena's fight program, while Wasp and a small group of special ops personnel would set up on planet as her support. One of the three shuttles would stay in orbit to provide a platform for both communications and information gathering. The other two shuttles would move out toward the Oort Cloud to see what could be learned about the mining operations.

"Ten minutes until orbit of Ephesus, ma'am," said the pilot of her shuttle.

"Roger," said Lee. "Inform the other two shuttles."

CHAPTER 67

MR. WATSON
(18 MONTHS AGO)

The evening after the failed raid on the cabin, Chambers walked into the *Travelers' Home*. The bar was empty except for one man: Anthony Moscatelli.

Chambers looked around the room, scanning for an ambush. There was none. His eyes finally settled on the young man. Anthony had a briefcase in his right hand. Chambers said nothing. Patience was a virtue when establishing control. Anthony was an unknown. Chambers had killed the man's uncle the night before. If Anthony wanted revenge, now was the time. Chambers had intentionally come unarmed. It was show of confidence, and yes, a gamble.

"Sir," said Anthony. "I'm afraid I don't know your name."

"I'm Mr. Watson," said Chambers.

"Mr. Watson," said Anthony. "Mr. Antonelli has left the premises. In fact, he has departed the planet."

"And you are here because...?"

"I understood you wanted to see the books for this establishment," said Anthony. "I have those ready," indicating the brief case by raising it slightly.

"Thank you, son," said Chambers. "First let's have a drink. Could you bring some sparkling water?" He looked around and said, "And where are the waitresses?"

"I sent them all home, each with a gift for the family," said Anthony. "I'll get the water."

Chambers drew upon 40 years of leadership experience to gain control of the operation. First, it was important to set the tone for the new leadership. Anthony had provided a list of operations that had been run by Mr. Antonelli. Most of these Chambers and Nemeth had already identified. Chambers was gratified that the list was complete, and surprised that some of the businesses on the list he had not been aware of.

Chambers, in his new identity of Mr. Watson, had Anthony put out the word that he wanted the leadership of each of the business to be standing by the next day for Mr. Watson's inspection. Anyone who didn't want to accept Mr. Watson as the new boss, would be free to leave the city prior to morning. There would be no retribution as long as all the assets were left undisturbed, and the person or persons were out of the city for good.

Anthony identified for Chambers those remaining who could be relied upon, and who would be problematic.

Chambers said, "Let's start with the problem children."

Anthony had picked the warehouse adjacent to the port facilities. The head man there was Giuseppe Mastromichalis. He was a short man, heavily muscled with a bald head and massive forearms. According to Anthony, he had been fiercely loyal to Mr. Antonelli. When told that Mr. Watson would want to meet with him and inspect his operation, Mastromichalis had reportedly said, "He's welcome to come; I can't guarantee he will be leaving so easily."

251

As Chambers and Anthony drove up to the front of the warehouse, Chambers noted the second story windows were open, and that no one seemed to be working in the yard.

The two got out of their vehicle and stood facing the open loading dock of the warehouse. In a moment, three men came out of a door adjacent to the loading dock. Mastromichalis was in the center, while two other men, one on either side, stayed a step back from their boss. Mastromichalis had a fierce look, his head was forward and his workman's shirt collar strained to contain his thick, muscled neck.

Chambers remained completely calm, waiting for the bull-of-a-man to reach him. Mastromichalis stopped a foot from the two, leaning forward to within a few inches of Chambers' face.

"Why are you here?" demanded the man.

Chambers answered slowly and deliberately. "I'm here to inspect your premises and your books."

Mastromichalis spat at Chambers' feet, raised his finger as if to jab at Chambers' chest. Before the finger could reach its target, the man's fist exploded in a spray of blood and bone. Mastromichalis screamed in agony and terror, staring incomprehensibly at the arm now ending in a stump at the wrist.

A fraction of a second later, the sound of a shot could be heard. The bullet that had taken the man's hand had traveled faster than the sound of the muzzle blast from Nemeth's sniper rifle.

Mastromichalis fell to his knees clutching his wrist and screamed, "Shoot this bastard!"

A moment later, Chambers heard two shots spaced about a second apart. He looked up toward the two open windows of the warehouse and saw snipers in each window fall backwards as Nemeth's shots plowed them rearward.

The two bodyguards stood frozen. They had expected to be able to take this old man, an interloper, with ease. The idea that the old man would have his own sniper in place before the meeting was simply not possible. The man on Chambers' right began to reach inside his jacket.

Chambers said calmly to him, "Best not do that," while wagging his index finger back and forth in a scolding motion. As he spoke, a red laser dot appeared dead center of the man's chest. The man froze and looked wide-eyed at Chambers.

Chambers looked to the man on the left, and said, "Let's get your former boss some medical attention."

The man responded in a rushed voice, "Yes, Mr. Watson," and he began running back to the warehouse yelling, "Medic! I need a medic!"

Chambers looked to the remaining bodyguard, who stood rock still as if waiting for his own head to explode. Chambers said to him, "Let's start with a tour of the warehouse, then the books. Please lead."

After that, there was no open resistance. Two of the dozen remaining properties were abandoned when Chambers arrived. Anthony set about moving people around and hiring new workers. At the remaining facilities, Chambers and Anthony were met by nervous but enthusiastic sub-bosses who seemed eager to accommodate the new boss.

CHAPTER 68

MR. WATSON
(PRESENT DAY)

Chambers, Nemeth and Anthony sat around a small conference table in Chambers' office. Chambers had moved his office from the Traveler's out to the port facilities. It was more central to the core of the business. Additionally, he wanted to set a different tone for his operations. Almost all the businesses were completely legal. He had shut down the abhorrent aspects of Mr. Antonelli's operations. No more drugs, prostitution, or protection operations.

Because he had returned the girls to their homes with the traditional compensation, Chambers was tolerated, even trusted, by the locals. He had monthly meetings over dinner with the local leaders, where he would hear their concerns. The locals were always treated with respect by Chambers and his people. Chambers had made it clear he would be very unhappy if he got news that any of his people had abused their authority concerning the local populace.

But one issue remained that he hadn't been able to crack: the mining portion of the operation remained somewhat opaque. It turned out that Antonelli had not really controlled the mining operation. He had provided the manpower and facilitated the orbital docking and resupply, but the mining itself was being handled by another part of the organization that didn't report to Antonelli. Because of this, the mining operation did not report to Chambers when he had taken over.

Chambers' efforts to move in on the mining segment had been rebuffed. Nemeth had been assigned the task of gathering information.

"Attila, please give us an update on the mining operation," said Chambers to Nemeth.

"Not much has changed since we last spoke," said Nemeth. "We can't get close enough with any vessel or probes. The defenses are sophisticated."

"Do we know what they do with it?" asked Chambers.

Nemeth shook his head. "No. None of it is coming back through our ports. All we can see is that there is a great deal of power output."

"Could it be similar to the generator of Alpha 51, perhaps? Are they using it for MWC projection?"

"Maybe," said Nemeth. "We don't have a scientist or the instruments to verify it one way or the other..."

"But?"

"It doesn't feel the same," said Nemeth. "Something else is going on."

"I agree," said Chambers. "Have you had any luck getting someone into the mining facility?"

"No," replied Nemeth. "Every time we send one of ours in as new staff, they either get rejected or we never hear from them again. Whoever is in charge is running a tight ship."

Chambers nodded and asked, "How are you coming with our fleet?"

Antonelli had left a small combination of space-capable vessels, used mostly for shuttling supplies, equipment and personnel to and from the orbital maintenance facilities. Among this hodgepodge of vessels was an old frigate that had been abandoned in orbit by pirates who limped into the system but could not keep it running. Antonelli had taken it over and used it to enforce tariffs on traders. It was over 100 years old, had no jump drive, and was in poor repair.

Chambers had asked Nemeth to have a look at the frigate to see if it could be repaired so as to be of any use.

"The vessels used for shuttling, storage and maintenance have been repaired so that they are at least safe to operate in orbit," said Nemeth. "The so-called frigate is coming along better than I thought it would when we first had a look at it a year ago. It now has a jump drive that, in theory, could be used within the system. Not enough to go anywhere else."

"Weapons? Shields?" asked Chambers.

"Shields are minimal, they won't hold up for long," said Nemeth. "I was able to get two older five-inch guns mounted and four relatively modern anti-ship missiles."

"That's good news," said Chambers with a rare smile. "Let's take a shakeout cruise to get the feel of it. We might need it at some point."

"We can go tomorrow, if you want," replied Nemeth.

Chambers said, "Let's do that. By the way, what did you name it?"

"The *Perseverance*," said Nemeth.

Chambers was thoughtful, and both Nemeth and Anthony waited patiently. Finally, Chambers turned to Anthony and said, "Anthony, can you bring us to date on our domestic operations?"

"Yes, sir," replied Anthony. "You have my report on expenses and sales. No real change there. Net profitability is steady. Your investment in the local community schools and hospitals is a drag on overall profitability. The locals are pleased..."

"And who is not pleased?" asked Chambers.

"Your sub-bosses," said Anthony. "It's not in the open, but there are rumblings. You've cut out the girls and drugs, and that was a major source of off-the-books income for them. They see the locals getting new things and it makes them restless."

Chambers smiled, nodded and said, "Thank you for that. I do need your honest views. Any word on our visitors?"

Visitors was their name for spies. Early in his tenure as boss, Chambers had surprised Anthony by simply ordering him to fire the spies they had uncovered that could not be turned. Anthony had asked, "*Why not kill them? Send a message to others who might try to infiltrate us.*"

Chambers had replied, "*Because there really is no reason to do that. There will always be spies, always. And we don't really have anything to hide, do we? Any spy will see that we are strong, and that is a good message to send back to their handlers. Sometimes the truth is the best deterrent. Also, let's not forget, we spy on them, and that is useful.*"

Chambers was aware that every organization would attract outsiders who would try to place spies inside his organization. His policy was to identify those spies and keep an eye on them. Eventually, he would try to turn them, get rid of them or maybe just ignore them.

"Mostly, just the usual," said Anthony. What he meant was that other crime families had sent spies in as workers to the various business. Anthony knew who they were and monitored their communications. That told Chambers who was interested.

Chambers looked at Anthony quizzically and asked, "Anthony, why haven't the other families moved against us? All they have done so far is send spies. They must know we killed some of their own when we took over. Aren't they duty bound to respond?"

Anthony shook his head, "No, sir. If you had been from one of the other families and did what you did, they would have had to respond with harsh measures."

"So, what do they think," asked Chambers.

"Mr. Antonelli is seen as a pathetic failure. You came out here with only two people and drove him out in one day," said Anthony. "No one, not even his own family, will follow up on that humiliation."

"Any chance he will lick his wounds and try to retake this operation?" asked Chambers.

"No chance of that," said Anthony. "His family disowned him after his failure to hold this place. Once that protection is lost, no one in his position lasts long. Too many grievances over too many years. He won't be back."

Chamber was thoughtful for a moment and then said, "You said, *'mostly just the usual.'* What's not usual?"

"There *is* something new," replied Anthony. "About a month ago, we got a couple; they came to work at the sporting arena. They are not from one of the other families."

"You're sure they are spies?" asked Chambers.

"Yes, sir. I asked Mr. Attila to have a look," said Anthony as he turned to Nemeth.

Chambers turned his gaze to Nemeth with a questioning look.

"James and Christina Wells. I watched them for a while," said Nemeth. "I now have them under continual surveillance."

"And?" Said Chambers.

Nemeth smiled, "They are from our old friends."

"Oracle? That's different. Are you sure?"

"I'm sure they are trained the same way," said Nemeth. "They use the same techniques. They are very good, really too good."

"What gave them away?" asked Chambers.

"First, they are too physically fit by a large margin. They can't really hide it. I arranged for one of the other workers to start an argument with the man and take a swing at him. It's the way he moved. He tried to hide his skill, but I saw it. Definitely trained by Oracle or someone who has been."

"And second?"

"And second, they are not really a couple," said Nemeth. "They act like it outside their living quarters, but when they think they are alone, not so much. Definitely not a couple."

"OK," said Chambers.

"How would you like me to handle it?" asked Nemeth.

"No need to do anything," replied Chambers. "We knew the Alliance would take notice at some point. Keep them under observation and try to find out who they are."

"You know they'll eventually send more," said Nemeth.

Chambers smiled and said, "Yes. Let them come."

CHAPTER 69

AN UNHAPPY MARINE

Corporal Baxter Jones was unhappy. He had joined the Alliance Marine Corps seven years ago. In training, he had always finished near or at the top of his class. Within a few years he had qualified for the elite strategic reconnaissance force and had served with distinction. About two years ago, he was told to report to his commanding officer at the company headquarters building. He entered the CO's office, came to attention, saluted and said, "Corporal Jones reports, sir."

His CO, Captain Jefferson, returned the salute, and said, "Corporal Jones, this gentleman would like a word with you." He nodded to his left where Jones noticed for the first time that another person was in the room. He was a tall man in civilian clothes, well over six foot tall, maybe 35 years old. His face was weathered, and he had the look of a marine who had spent long years in the field.

Then Captain Jefferson stood and walked out of the room, leaving Jones alone with the man.

"Have a seat," said the man, indicating a chair across from where he was seated.

Not knowing what else to do, Jones took the seat as ordered. He stifled the urge to ask, *"Who are you?"* and simply waited for the man to speak.

The man looked at Jones closely with a questioning look on his face, as if by looking he could see something of value or interest. Jones looked back. He didn't know who this man was. Probably a superior judging by the age and his deportment. But Jones didn't really like getting jerked around. He decided to wait and see what the man wanted.

Finally, the man said, "Well, you just passed the first test."

Still, Jones didn't answer. He didn't like the man's tone and he could tell the man expected him to say, *"What test?"* He was beginning to get angry.

After a moment, the man said in a serious tone, "I'm Mr. Black. I represent an arm of the Marine special operations forces. I'm here on a recruiting mission."

Still Jones said nothing. No question had been asked, and he stubbornly remained silent.

The man nodded and continued. "Have you heard of Oracle?"

"Yes," replied Jones simply. "Rumors only."

Everyone had heard of Oracle. It was a near-mythical counter terrorism organization that supposedly included both Navy and Marine special operations personnel. It was a deep cover, black op organization. Jones didn't elaborate. He had never had any official information in his training about Oracle. And also, he knew that whatever he thought he knew was likely to be wrong.

"We've looked at your record, spoken to your chain of command. We think you might be a good candidate. Is this something you might be interested in?"

"I'm a Marine and I'll take whatever assignment I'm given," said Jones. "If you are asking me to be a volunteer, I'm happy to learn more about it before I would agree."

"I think that is wise," said Mr. Black. "The training lasts two years. We only recruit Marines like you, who have been at the top of your training and have solid recommendations from your superiors. Even from that pool, about 90% wash out. Of those who fail, most quit, some are injured, some die. Our casualty rate in field on operations is higher than the norm for combat."

"I have a mother who is living. Will I see her?" asked Jones.

"Yes, but not often and not at predicable intervals," said Mr. Black. "We will give you a cover assignment that you can use to explain to family and friends what you're doing."

"Will I be undercover?" asked Jones.

"Going undercover is part of our mission, yes," said Black. "But not everybody does that. *Everyone* is trained in the full spectrum of missions, but our focus is counter-terrorism strike operations."

"OK, I'll do it," he said simply.

The training was horrible, and yes, almost all of the recruits eventually washed out; some were injured and couldn't continue, but most just quit. Jones didn't see anyone die, but it really was that hard.

Finally, he had finished. He was overjoyed to be assigned to an operational detachment. His team leader, First Lieutenant Theodorus Wasp, was a legend in the Corps. He had been the youngest sergeant major in the history of the Marines before going to OCS and becoming an officer. He had a reputation for leadership and competence in combat. Wasp had led the mission on Alpha 51, in which his team had inserted, by moo suits for God's sake, launching from a passing asteroid.

CHAPTER 70

MR. AND MRS. WELLS

But almost immediately, Jones' dive into Oracle had gone sour. His team had launched out on a mission with almost no preparation right after Wasp had gone to see the Marine Corps Commandant, General Motubu. The mission was to hunt down a missing naval shuttle that had been on a classified mission, recover the shuttle's crew if possible, and assist with whatever mission that team had been on.

But then Wasp had assigned Jones and Sergeant Janice Wallace to undercover duty. They were supposed to be a married couple immigrating to Ephesus looking for work. Sound exciting? No, it wasn't. Jones thought Wallace was a bitch, and that was all there was to say about that. She was ten years older than Jones, but didn't look it, so they were plausible as a couple at first glance. She was not happy to be stuck with a newbie on an undercover assignment. She had done undercover work before, and she was good at it. And, as she made it clear to him, he was *not* good at it.

She made a point of correcting him when they were alone at night. She could be scathing; always saying things like:

"You need to stop looking around all the time. Damn it, you look like you're spying." And

"You're supposed to be my husband; act like it in public. No one is going to believe us if you keep deferring to me like a puppy dog." And

"Do you even remember your cover identity? You told that pinhead you work with you were from Indiana. *Indiana!* That's where *you*, Corporal Jones, are from, but Mr. Wells *isn't from Indiana!* He's from Oklahoma. There is a difference."

The last straw came when Jones got into a fight. A rude Italian guy had been hitting on Wallace in front of Jones. She, acting the part of a married woman and not the part of a highly trained special ops warrior, had done a good job of pretending to be flustered and embarrassed at the attention.

Jones was pissed off and stepped in to confront the man. The guy had immediately, and somewhat unexpectedly, taken a swing at Jones. Without conscious thought, Jones had parried and then used a submission hold to calm the guy down. He saw the look on Wallace's face, and he knew he had screwed up.

Later that night she had unloaded on him. "You idiot! Anyone watching that who knows jack about combatives will know you're not an out-of-work oil worker."

"What did you expect?" said Jones. "He swung at me."

"I expect you to take the punch, or to do whatever is needed to keep our cover because, damn it, *our lives depend on it!*" Wallace was furious. "You do know the target is Jay Chambers, right? He was one of us, one of the best of us. If he's got eyes on us, he'll know you're Oracle trained."

"But the guy was messing with you," said Jones. "I'm supposed to be your husband. Wouldn't a husband come to your defense?"

"Idiot!" Wallace spat the word. "Did it ever occur to you that the fight might have been a setup? A test? That someone wanted to see how you would respond? By all means, get into it with the guy, defend my honor. But don't show off your martial arts skills."

Just then her handheld device gave off a warbling tone.

"Shit, that's Wasp," she said.

She activated the device and they both saw a miniature image of Wasp projected above the screen, which she held horizontally.

"Yes, sir," said Wallace.

"I got your digital report, thank you," said Wasp. "Just wanted to let you know we are sending Lee to you in the next 24 hours. The plan is for her to enter the arena complex and work her way up until contact is made with Chambers."

"Roger, sir," said Wallace. "We'll be ready. I do need to report that there was an incident and that we may have been compromised."

Wasp showed no sign that he was surprised or alarmed. "Tell me."

"One of the workers started in with me," said Wallace. "My faithful husband stepped in, and a punch was thrown. Jones handled it, but we're thinking it may have been an intentional provocation to see how Jones responded. He may have been a bit too adept. My fault, I should have anticipated this and prepared the both of us better."

"No doubt," said Wasp. "As the senior person, you are always responsible for the success or failure of the mission. But the damage isn't likely to be fatal to your mission. Our best information is that Chambers and Nemeth are the good guys. I fully expected them to sniff you out eventually. In fact, this is a good thing. If you're blown and you haven't been rolled up yet, it reinforces the idea that Chambers is on our side."

"Your orders, sir?" asked Wallace.

"Make contact with Lee," said Wasp. "Get her on the fight roster, but do it in a way that doesn't trace back to you two. She'll have to make her way up the tournament ladder on her own merit. For now, keep doing what you're doing. Be prepared to assist her if the situation warrants."

"Roger, sir," said Wallace and the link was terminated.

Jones looked at her and she said, "What?"

"You defended me," said Jones.

"No, I didn't," she responded. "Wasp isn't a wet-behind-the-ears lieutenant. He was an NCO, a CSM for God's sake, and a good one. He knows what happened. But he is right. I am responsible for this mission. Your failure is my failure."

Jones looked at her and began to blush as he realized his mistake with the fight had made her look bad. He saw now that by arguing with her, he had just been trying to find his way out of a foul up he had created.

He said in a low voice, "I'm sorry I screwed up, Sergeant. I have no excuse."

She looked hard at him and then broke into a smile. "Actually, it was pretty cool what you did. Thanks for sticking up for me. Another time and another place, that would have been just right. You stuck up for me, your wife, and you didn't hurt the poor bastard. Someday, you are going to make someone very happy. I just wish you had taken it down a notch."

"What now?" he asked.

"Now we get ready for Captain Lee," she said. "If you think I'm a bitch, you haven't seen anything yet."

He smiled and said, "Really? Worse than you?"

"Damn right," said Wallace smiling. Then with a serious tone she said, "You heard those pirates we interrogated. She took them out while she was unarmed, and she used their own weapon to emasculate one of them. You saw how frightened they were when her name was mentioned. They weren't afraid of us, even though we were pointing guns at them. They were afraid she would come back."

CHAPTER 71

MARGARET THE
MAGNIFICENT

Corporal Jones and Staff Sergeant Wallace watched from a building opposite the coliseum as Mei Ling Lee walked toward the back entrance. They were set up as a sniper's nest with Jones on the rifle and Wallace on the spotter's scope. They had a clear line of fire and they were listening via a concealed device in Lee's ear. Captain Lee had made clear their job was to shoot only if her life was in danger.

This was the entrance that the competitors used. Lee had met briefly with the Wells. They had gotten her an appointment with the fight scheduler, but hadn't been able to clear her through security, which would take a few more days. Lee had said she didn't want to wait, and she hadn't.

In front of the entrance were two guards, big men. Lee had learned from the Wells that the men were former fighters. These men had, at some point, washed out of competition, and since they had no real skills needed in the workplace, they had gravitated to being bouncers. Their job was to turn away unauthorized persons, most of whom were trying to see their favorite fighters up close. Others were just looking for a way in without paying for a ticket. According to the report from the Wells, the two were bullies who occasionally beat people if they thought they could get away with it.

As Lee approached, the two men blocked her way. The one on her right was a huge man, well over 300 pounds, whose muscle had gone to fat over the years since he had been in the ring. The one on the left was younger, maybe 30. He was early in the process of going to seed, but still looked relatively fit.

The fat one said, "Wrong entrance, missy."

Lee looked up at him, her cold face on. "Not missy, my name is Margaret. I'm a fighter and this is the right entrance."

"I know all the fighters; you're not on the list, so piss off, bitch!" The man used his palm in a thrusting motion, as if to push her back.

Lee moved quickly. She trapped the extended arm with her forearm. Using her free hand, she struck with her palm and smashed the man's arm at his elbow, hyper-extending it and causing an audible snap as the bone dislocated. The man screamed and tried to slap at her across his body with his other hand, but he couldn't reach. Then Lee slammed her foot down on the side of his knee. Again, the sound of snapping tendons and bone. The man was screaming in earnest now, crying. He fell to his knees.

With his face now level with Lee's, he gasped, 'What the hell, lady?"

Lee placed her finger on the man's eye and turned to look at the other guard who had begun reaching for his weapon under his jacket.

"Draw your weapon and I'll take his eye," she said calmly. The man dropped his hand to his side.

Turning back to the man she said in a cold but steady voice. "You called me 'missy' and 'bitch'; those are not my names. What is my name?"

"For God's sake, it's Margaret," said the man.

Lee slapped him hard across the face and pulled him close. "You showed me disrespect. From now on when you see me, you will address me as Margaret the Magnificent. Now say it."

"Margaret, the Magnificent," said the man in a whisper.

Lee slapped him again. "No, no. That won't do at all. Please say it louder, like you mean it."

"Margaret the Magnificent!" shouted the man.

"Much better," said Lee. She let go of the man who fell to his face, sobbing.

Lee turned to the other guard. "I have an appointment with Mr. Chang, the fight scheduler. Please show me the way."

"Holy crap," exclaimed Jones from their snipers' nest across the street. "I've never seen anything like that."

"I told you," Wallace said.

"You said she was a bitch, not that she was El Diablo," said Jones. They watched as she followed the guard into the building.

"She's on her own for now. Let's hope she can handle Chang," said Jones. "He's ornery. He could have her thrown out or arrested."

"He could try. But I think she'll be fine," said Wallace. "Margaret the Magnificent has persuasive skills."

CHAPTER 72

MR. CHANG

The guard who was now serving as Lee's escort walked in front of her down the narrow corridors of the coliseum. Word spread fast and people stopped to stare at Lee and scurry out of her way if they got too close.

The man stopped and knocked on a door to an office overlooking the coliseum. He opened the door, stuck his head in and said, "Mr. Chang, a fighter to see you, sir."

"Who is it?" A voice snapped with a heavy Chinese accent.

"It is..." The man paused and looked back at Lee. She raised her eyebrows and tilted her head as if to say, 'Well, you know my name.'

"Margaret... the Magnificent," he finished with emphasis.

"Well, if she is magnificent, please send her in."

Lee entered the office. It was a surprisingly large room, with an interior window that looked out over the coliseum and had a clear view of the fighting cage at the center. Obviously, Chang was considered an important person to have such an office, and to have the guards address him as Sir and Mr. Chang.

The two stood staring at each other. Finally, Chang said, "Yes?"

"I'm a fighter; I need a match," said Lee.

The man sat down, opened a large notebook, ran his finger down the page and said, "Hmmm... you have no record, you are new. I can start you at the bottom. First match in a month."

Lee said, "That won't do; I need a match tonight."

Chang looked up at her with wide eyes. He said with sarcasm, "Well, Ms. Magnificent, we have rules. And if you think..."

He never finished his sentence, because in that instant, Lee was next to him leaning over his shoulder looking at the scheduling book.

Chang jumped up out of his seat and took a step back from her. "What the... Are you a witch? How did get across the table?"

Lee looked at him with the cold face and said, "Whatever else I am, Mr. Chang, I am a fighter."

She turned back to the book and said, "Here, you have an open slot. Put me against this fighter," pointing to a name on the page.

"No, no," said Chang. "That is not for you. This is Won Tae; he is not for you at all. He killed the last man he fought against. The berth is empty because no one will fight him. In the arena there is no referee, nothing to stop one fighter from attacking a downed opponent. Won Tae has a temper. He would crush one such as you."

"I see what you mean," said Lee. "He would not be for me at all. Totally unfair. Better to let me fight Won Tae and another at the same time. That way it might be an even match."

"Are you insane?" asked Chang. "Mr. Watson would fire me in an instant if I allowed a tiny woman such as yourself to be killed in the arena."

"I see your point, hmmm..." said Lee thoughtfully. "Let me ask you this: do you make bets on these fights?"

"Me? Of course not! That would be a conflict of interest," said Chang clearly trying to muster indignation.

"Of course," said Lee. "But you know people who make bets, and they could help you place a bet, off the record, yes?"

Chang looked intensely at Lee. She had struck a nerve. "What another person does with his own money, that is his own business."

"I understand," said Lee. "Here is what I suggest: Have your friend place a bet for me to win against Won Tae and any other fighter you choose. I assure you I will win. Your friend will be wealthy and you as well. What are the odds, a hundred to one?"

"Three hundred to one, easily," said Chang. "You cannot win such a fight. It would be suicide for you and murder for me."

Just then a knock came at the door and a bookish looking man in a white collared shirt poked his head in. He said to Mr. Chang, "A moment of your time, sir."

"Yes, come," said Chang waving the man forward. The man stepped into the room, saw Lee and hesitated, clearly unwilling to step any closer to her.

"What is the problem?" shouted Chang. "Are you afraid of this tiny woman?"

The man was trembling and said in a weak voice, "Yes, sir. I am afraid. Chun Gun confronted her at the gate, and she broke his arm and leg in an instant. She made him squeal like a pig. Gun has been taken to the hospital. I have come here to warn you about her."

Chang stared at the man and then looked at Lee. He turned back to the man and said calmly, "Thank you, Mr. Kwon. That will be all for now."

The man left in a hurry, clearly relieved to be out of the office and away from Lee.

"You injured my man at the door? What possible cause did you have to do that?" Chang asked.

"He was disrespectful to me," said Lee. "I told him my name was Margaret and he instead called me missy and bitch, and then he tried to lay hands upon me. He is fortunate: his wounds will heal, and he may, God willing, still have the ability to have children to carry his name. Others who showed lack of respect were not so lucky."

Chang faced Lee and was about to speak when suddenly she was very close to him, inches away. He had not seen her move. How had she done that? He began to squirm.

Chang said in a shaky voice, "Margaret the Magnificent, you are on tonight in the first match. You will fight Won Tae and Jin Do. I pray for your survival."

"Thank you, Mr. Chang," replied Lee also in a formal tone. "I trust your prayers will be answered."

Mr. Chang visibly showed relief that the confrontation was not going to end in violence.

"One more thing," said Lee.

Chang tensed and managed, "Yes?"

"After tonight, you will be a wealthy man because of me," said Lee. "You will be in my debt."

CHAPTER 73

LADIES AND GENTLEMEN!

The arena was packed with excited, cheering spectators. Word had gotten out that a mysterious new fighter had insisted on fighting the infamous Won Tae, who had recently killed another opponent in the ring. So confident was this fighter, Margaret the Magnificent, that she had also insisted that she fight another along with Won Tae.

Lee moved through the crowd toward the arena, which was a simple concrete square slab, set off by a ten-foot-high chain fence. The coliseum had provided her two uniformed guards to help push her way through the throng of excited people. They made a halting forward progress, but Lee kept getting buffeted right and left. Many seemed to want to just touch her, as if she were a religious figure.

At one point there was a scuffle in front as someone pushed the guard. The guard snapped out a telescoping baton and struck his attacker on the thigh. The man went down right in front of the guard. There was a confusing delay as the guard tried to move the prone man out of the way.

While this disturbance had the attention of all, someone pushed hard against Lee's right side. She moved away from the pressure, and as she did, she felt a sharp sting in her left shoulder.

Lee spun toward the person and yelled, "Back off, asshole!"

The man who had made contact with her was none other than Mr. Chang, the fight scheduler. He raised his hands in a placating gesture and said with a smile. "I won't owe you anything, Ms. Magnificent."

In a moment, Lee was swept forward toward the cage, and she lost sight of Mr. Chang.

As she approached the wire cage, she could see there were a dozen or so people blocking the path she would need to take to get to the only opening on her side of the platform. Not wanting to wait until it was cleared, she broke from her formation, jumped the last four feet to the fence, and quickly climbed it. At the top, she flipped head over heels, dropped the ten feet to the floor inside the cage, and landed easily on her feet.

At this little bit of showmanship, the crowd went wild, and began chanting, "Mar-gar-et, Mar-gar-et!"

Lee looked around the enclosure. Nothing special: a square concrete slab contained by 10-foot-high fencing on four sides. Two sides opposite of each other had a door which could open. She suspected the door would be locked from the outside until the match was over to prevent any fighter from escaping if the fight didn't look survivable.

Lee felt a twinge of nausea and a slight vertigo. She recalled the slight pinch she felt when Chang pushed into her and his words, 'I won't owe you anything, Ms. Magnificent.'

She fought back a rising panic as she realized Chang must have injected something into her left arm. She slowed her breathing and ran through the possibilities. Some sort of toxin, surely. She tried to feel her body: what was working, what was impaired. The nausea passed quickly, but there was still some vertigo, but that too was fading. A narcotic? No.

She didn't feel the euphoria that might bring. Then she noticed a tingling in her left arm.

There it was: a paralytic. The toxin was designed to impair or immobilize her left arm. It was probably short-lived and would be out of her system quickly. Chang had obviously bet against her, and the toxin was designed to get Lee killed before it wore off. By the time of any autopsy, the toxin would be out of her system.

Lee tried to shake her left arm. She could raise it a little, but not enough to use it to block. She scanned herself for other symptoms. The nausea and vertigo had subsided, she assumed that was a side effect from the injection, which had now passed. She shuffled her feet, and realized she would need to compensate for the lack of participation of her left arm.

Luckily, the poison didn't seem to impact her lower extremities. Her legs moved fine. She tried a kick in the air. It worked, but she would still need to account for the lack of counterweight the left arm usually gave her.

Just then she realized she had been through this before. Guyuk had trained her hard for a month of fighting with one arm strapped tightly to her torso. Had he somehow known this would happen? He did seem to have visions about future events. He had seen her crash land in the meadow 30 years before it happened. What had he said to her during training for one-arm fighting? 'Daichi Tengri, there are dark times ahead. I see treachery in your path. It is best to be strong.'

All right then, thought Lee. Let's do this thing. Guyuk had given me this gift, and I'm not going to waste it.

Lee was brought back to the present by the roaring crowd. Won Tae was entering the cage. The crowd seemed equally split between cheers and boos. Won Tae was a giant of a man, six foot five at least, and 300 pounds of solid muscle. His face was tattooed with some sort of dragon motif that started at his cheekbones and continued down his neck and on to his bare shoulders.

As he entered the ring, he ignored Lee and turned to the crowd, raising his fists above his head in a gesture of dominance and confidence. The crowd grew even louder in its combined approval and condemnation.

Behind him, unremarked by the crowd, entered the man Lee assumed was Jin Do. He was smaller than Won Tae, maybe six feet, but heavily muscled. He made no gesture to the crowd, but simply began to warm up with a series of punches and kicks to the air. Watching him move, Lee realized he would be a significant opponent. She would have to take that into account; couldn't ignore him for a moment.

She thought about a strategy. She had options. She could try to drag it out, hoping one or both of them would tire so she could take advantage. Or she could try to take one of them out quickly so she could deal with the other.

She didn't get the chance to do either.

"Ladies and Gentlemen," sounded the disembodied voice of the ring announcer. "Tonight, we have a classic match of David versus Goliath. Or, in this case, two Goliaths. The Challenger, hailing from parts unknown, at five foot two inches tall, weighing in at 117 pounds, with zero wins and zero losses, Margaret the Magnificent!"

Lee raised her good hand and turned in a circle to face the crowd. The crowd exploded in cheering. Lee could see they were on the verge of an uncontrolled riot. Some were rushing the cage and being pushed back by overwhelmed security guards.

"And the defender and current champion, hailing from Earth, at six feet five inches, weighing in at 290 pounds, with 15 wins, all by knockout, Won Tae!"

Tae raised his fists above his head, shaking them at the crowd. Again, the spectators exploded in a confusing mixture of cheers and jeers.

"And joining Won Tae against the challenger, hailing from Epsilon Five, at six feet zero inches, weighing in at 200 pounds, with two wins and five losses, Jin Do!"

At this the crowd booed in earnest and together. Apparently, they either did not like Jin Do, or they didn't like the idea that he would join in the fight against the diminutive Margaret.

Jin Do raised a single hand in polite and modest recognition of the crowd. His polite gesture earned him even greater scorn from the crowd. He nodded as if unperturbed.

"Ladies and gentlemen," continued the ring announcer's voice. "By the rules of tonight's match, only one fighter may be declared the winner. Should Margaret be vanquished..." Intense booing from the crowd at this. The announcer had to wait for it to die down to continue. "Won Tae and Jin Do will need to settle between themselves which is the sole remaining champion. As always, ladies and gentlemen, there is no referee for tonight's match. A competitor may tap out at any time and concede the match to his or her opponent."

The crowd exploded again at this. Lee heard shouts of 'No prisoners,' 'No retreat!' And 'No tapping out!'

This is a bloodthirsty audience, thought Lee. She knew that Won Tae had killed his last opponent, and she wondered if the unfortunate man had tried to concede. Lee guessed there was no penalty for continuing to attack an opponent who tried to tap out, despite the rules to the contrary.

The three faced each other, Won Tae and Jin Do side-by-side, both facing Lee. They dwarfed her, and suddenly the crowd became quiet as if they had just then realized the seriousness of the match. The tiny Margaret was surely about to die at the hands of these men.

CHAPTER 74

A NIGHT TO REMEMBER

"Fighters, stand by..." Both Won Tae and Jin Do moved into a fighting stance, fists up. Lee remained as she was, hands at her side seemingly unperturbed by the threat now facing her. "Fight!" Came the command.

Immediately, Won Tae turned and struck Jin Do with his fist on the side of his head. Jin Do, who was completely focused on Lee, had not seen the blow coming. Lee watched as Jin Do's jaw clearly dislocated, teeth flying out of his mouth. The man's head rocketed to the side, and it was clear that Jin Do was unconscious before he hit the ground. He rolled once and stopped, face down, in what became a rapidly spreading pool of blood coming from his ruined mouth. Lee could see he was choking on his blood and would suffocate in minutes if not given care.

The crowd gave a collective "Ohh!" at the unexpected assault. This was followed by booing.

Clearly, Won Tae felt he could handle Lee on his own, and wanted to remove Jin Do easily and right away while he was unsuspecting of an assault.

Lee made no move. She had learned long ago in her martial arts training to never show surprise, even when the opponent took an unexpected action.

But she did start making a clicking noise with her tongue.

Won Tae smiled, looked at Lee, and said, "Now you, Missy Bitch."

Lee said nothing. Clearly, Won Tae knew that Lee had humiliated the guard at the entrance for calling her Missy and Bitch. His message was clear: You won't punish me for misusing your name.

Lee sprang forward, raised her right knee and unleashed a thrust kick directly into Tae's ribs. He made no effort to block or move out of the way, and the force of the kick knocked him back several feet. The sleep-making technique taught her by Guyuk had apparently worked.

Before Tae could recover, Lee immediately went to the unconscious Jin Do, and quickly moved him into the recovery position, on his side with one arm extended. It was all she could do, and she hoped it would open his airway enough to keep him from choking to death while he lay there.

She would have tried to clear his airway, but Tae descended on her with a flurry of kicks and punches, forcing her to back away from the unconscious man and to defend herself.

The crowd went wild when it realized Lee had moved to save her opponent's life, even in the face of risk to herself. 'Mar-gar-et! Mar-gar-et.' It shook the coliseum.

Lee was on the defensive, but she knew, at least for now, she had the advantage. Tae was clearly furious that she had landed the kick and that her act of helping the downed Jin Do had triggered the crowd in her favor. She had used the sleep making technique to get that kick in. But she was not sure she could use it again, because the noise from the crowd would drown out any clicking sound she tried to make.

Tae was fast and strong, he came on relentlessly, maybe even a bit recklessly. As she backed up, dodging his punches and kicks, she feigned a slight stumble, trying to draw him into a full commitment.

It worked. Tae lunged for her, arms outstretched clearly hoping to grab her and use his weight advantage to take her to the floor. At the last moment Lee shot out a defensive sidekick striking his exposed rib cage. It was, or should have been, a fight-ending blow. His weight, magnified by his forward momentum, took the full force of her kick. It should have crushed his rib cage and sent him to the floor and then to the hospital.

It didn't. The blow did knock him backward a step, but he didn't go down. Lee couldn't understand it. She had felt the solid connection of the kick, felt the ribs give way. Was Tae on some sort of stimulant that caused him not to feel the pain?

Then she saw it. From Tae's lower torso, she saw a piece of bone protruding from his rib cage. But something was off. The bone wasn't white, it was black. Then she knew: Tae had been surgically enhanced for fighting. Some sort of cybernetic modification. Her kick had broken his ribs, but those ribs were metal or some sort of synthetic composite. It was likely that the modifications increased his prowess as a fighter and made him inured to pain.

Before she had time to think about it, Tae was on her again. Lee was in full defense mode. She needed space and time to decide how best to fight a man who could not feel pain.

Lee went for his legs, hoping to take out his support even if he couldn't feel the pain. She kicked hard at his thighs, knees and shins. Nothing happened. He smiled at her, a knowing smile. He knew that she knew he was virtually invulnerable and that he would not stop until she was dead. No tapping out would save her.

For the first time ever, Lee began to feel the tingle of panic. She had faced death before, more than once. But on those occasions there was always a glimmer of hope. Now there was none.

Lee decided she would go down fighting. If her death was inevitable, she would make sure there was a cost to her opponent, even if he couldn't feel it.

Tae came at her head on. Through the soul sliding technique that Guyuk had taught her, she saw that Tae intended to grab her in a bear hug, to crush her spine.

She let him take her. But before he could close his arms around her, she raised her good right arm above her head. As a result, when he pulled her into him, that arm was free.

As he pulled tight, she went for his eyes. She put her thumb against his left eyeball and pressed inward with all her might. It was a move designed to blind an opponent, and she would normally not use it in a match. But she realized she was fighting for her life. As she expected, the flesh gave way as her thumb penetrated the eye socket. She thought he would scream and push her away, but he didn't.

In a horror of realization, she saw that his eye was not human flesh, but rather some sort of cyber-mechanical construction. It protruded from the eye socket, dangling from a tangle of wires.

The crowd went silent, as they struggled to understand what was happening. Then there were gasps, and shouts of "robot!" and "foul!"

Tae was not deterred. He spoke so only she could hear, "I don't need to see you to kill you, little bitch. And I'll have a new eye by tomorrow."

Lee could feel the incredible strength of his grip, crushing the air from her lungs and placing great stress on her rib cage. It was hopeless, but still she fought, ripping the flesh from his face revealing the metal underneath. He was grotesque, and the crowd let out a collective "Oh!" She pounded his head with her fist and struck with her knees up into his abdomen.

Lee was beginning to lose consciousness. She knew this was the end, that she would die. All that mattered now was that she went down with the dignity of never giving up.

Just before she knew she would lose consciousness, she felt a heavy jolt, throwing her backward. She landed on her back with Tae on top. He loosened his grip on her, and she struggled desperately to get out from underneath his weight.

As she scrambled free, she saw what had happened. Jin Do was back on his feet, his mangled, bloodied jaw hanging loose from the rest of his face. He had apparently regained consciousness and attacked Tae from behind. *Must have kicked him*, Lee thought. Tae rose up, his own mangled face a horror to see, and turned toward Jin Do.

As he stepped toward Jin Do, Lee leaped onto Tae's back, placed her right palm on his chin, and the near-useless left forearm on the back of his head and twisted. His head rotated beyond 90 degrees and Lee heard a distinctive snap as his spine broke. Tae fell to the floor and twitched for a few seconds, then lay still. Apparently not all of him was metal. His spine was bone, and Lee had just splintered it.

The crowd cheered, but the fight wasn't over. Lee turned to face Jin Do. He was no match for her now. He had risen to the occasion and delivered a great blow to Tae, but now he could barely stand.

He raised his fists into a guarding position, and locked eyes with Lee. Then he nodded, went to one knee, and tapped the concrete with his palm to show his capitulation. The crowd went dead silent in anticipation of what Margaret would do.

Lee approached him, pulled him to his feet, hugged him and then took his hand and raised it with her own to show the crowd that she considered both of them to be victors.

The cheering was something to behold. Lee had saved Jin Do from almost certain death, and he had risen, as if from the grave, to save her. Then, in an act of gracious defeat, he had conceded. She, in her turn, had acknowledged their co-victory. The crowd went ballistic.

CHAPTER 75

SAFE HOUSE

Lee woke in an unfamiliar place. It was dark, she was lying on a simple cot with a blanket pulled up across her shoulders. She could hear low voices in the next room. Someone was intentionally being quiet so as not to wake her.

Then it came back to her, the fight with Won Tae, the crowd going crazy. At first, she could see no way out of the coliseum; the crowd was simply insane. If she left the cage, she felt like she would be torn apart, even by her enthusiastic fans, who were again chanting, "Mar-gar-et! Mar-gar-et!"

Lee had begun to feel ill, perhaps from the poison shot in her arm, or maybe the injuries during the fight, or maybe just the overload of adrenaline rushing through her system. Probably all three.

Then a miracle of sorts happened. Uniformed guards entered the cage and surrounded her in a protective circle. They looked different somehow from the relatively shabby ones who had escorted her into the cage. To her amazement, she saw it was Wasp and five of his team dressed as coliseum security. Once again, Wasp to the rescue.

Lee approached him, opened her arms as if to embrace him. Her mind at the limit of what she could process, she felt a strong vertigo and nausea, and she worried she was about to throw up on her rescuer. Then nothing.

Now she was in this room. She spoke, "Hello?"

A moment later Laura came in, a warm smile of her face.

"I don't think I've ever been so happy to see a doctor," said Lee.

"How are you feeling?" asked Laura.

Lee thought about that. She tried to lift her left arm, and it did move more easily, not the full range of motion she was used to, but better than the almost nothing she had during the fight. She moved her right arm and felt, and then saw that an IV had been started there.

"The arm is getting better," said Lee. "Did you give me something for that?"

"Yes," said Laura. "It's a good thing Wasp and his team got you out of there when he did. That toxin would have been out of your system, but the damage to your nervous system would have been permanent."

"Where are we?" asked Lee. "And how long have I been out?"

"A safe house Wasp set up for the team," said Zakany. "We're about a mile from the coliseum." Laura looked at her watch and said, "You've been here 12 hours. You've been out for a little longer than that."

"What about Jin Do?" asked Lee. "He was hurt bad. Did he get out? Get medical care?"

"Wasp's team got him out OK," said Zakany. "Took him to a local hospital. Even left a guard with him to make sure there were no shenanigans."

"Thank you," said Lee. "Is Wasp here?"

"He's been in and out," said Zakany. "Checking on security. Emotions were high, mostly on your side, but Won Tae had his supporters, and they are apparently very upset with you. Sergeant Johnson and your support team, the Wells, have been here the whole time."

"Do I need this thing?" Lee gestured to the IV bag on her right arm.

"No. That can come off," said Laura. "You will need time for the left arm to fully recover. Your ribs are bruised. I suggest you don't fight a 300-pound cyborg for a few days." She smiled and said, "Seriously, give it a rest for now. You have tissue and nerve damage. Working it too hard too soon could make it worse."

Lee sat up and could feel the soreness in her ribs. Laura came around to her right side and disconnected the IV that had been inserted into her arm. She placed a bandage on the place where the needle had been inserted and said, "There, almost as good as new."

Lee stood up and walked unsteadily out of the bedroom and into the common area. Sergeant Johnson was there along with Corporal Jones, Sergeant Wallace and another Marine from the team. They all stood, and Johnson said, "Ma'am, I hope you are feeling better. That was one hell of a fight."

"As you were," said Lee. "Thank you and your team for getting me out of there. I'm pretty sure I couldn't have done it myself."

"Happy to do it, ma'am," said Johnson. "The Wells made it possible," she looked at Jones and Wallace. "They got us the uniforms and the clearances."

Lee looked at the two and said, "Well done, and thank you." They both nodded at the praise. "I take it your cover is blown?"

"Yes, ma'am," said Wallace. "We had to pull out all the stops to get Lieutenant Wasp and the team into the ring as security to get you and Jin Do out. We won't be able to go back."

"What about Chang?" asked Lee. "He poisoned me before the match."

"He must have bet against you," said Wallace. "Apparently, he used someone else's money to place the bet, because he was found dead at his desk after the fight. His throat was cut."

Lee nodded and asked, "Can you give me a quick update?"

"Yes, ma'am," said Johnson. "There was some rioting and looting around the coliseum after the fight. Lieutenant Wasp set up security for this building, including a sniper team across the road in front. We've been rotating our personnel through on four-hour shifts to maintain alertness. Things have calmed down a bit. The test will be nightfall, if the rioting starts up again."

"Do you think that will happen?" asked Lee.

"Probably not," replied Johnson. "The earlier disturbances seemed to be fueled by the emotions brought out in the fight. We're thinking that will have calmed a bit by tonight."

"Any sign Chambers has taken note?" asked Lee.

"No official word, though he could hardly have missed it," said Johnson. "Multiple videos of the fight are out on the local internet. Your face is everywhere. No way he could have missed that you are here."

Just then a voice squawked on Johnson's radio. She held her wrist microphone to her mouth and said, "Go for Johnson."

"Sergeant, this is White at the front. There are two men at the door," said the voice. "They asked to see Margaret the Magnificent."

"Have you ID'd them?" asked Johnson.

"Yes, Sergeant," said the voice. "Retinal scans confirmed: Jay Chambers and Matthias Nemeth."

CHAPTER 76

A REUNION OF SORTS

Sergeant Johnson looked at Lee with a questioning look.

"Escort them up," said Lee. "Let's get Lieutenant Wasp back here, please."

"Yes, ma'am," said Johnson and she turned away and began speaking into her transmitter.

Lee turned to the other special operations Marine and said, "Where is Raymond just now?"

"Ma'am," said the Marine "Mr. Raymond is out doing a recon of the mining operations. We can connect to him by video if you need him."

'Yes, let's set that up," said Lee. "Let me know when he's ready. Where is Danner?"

"He's in the command shuttle in orbit, ma'am. We can connect by video or have him here in an hour."

"Video will be fine," said Lee.

Lee turned to Zakany who appeared stunned. Her face was flushed, and she was looking into nothing as if she were in a trance.

"Laura," said Lee. "Are you going to be OK?"

"What?" said Zakany. Then she turned to see Lee and her eyes widened. "The father of my child, my husband..." she trailed off.

"Laura, take the time you need to get reacquainted," said Lee. "It's going to be a while before we can get everyone online."

One the Marine guards came into the room and said, "Ma'am, Admiral Chambers and Master Chief Warrant Officer Nemeth are here." He stepped aside and in walked Chambers with Nemeth behind him.

To Lee, Chambers looked older, but still fit. He was dressed in casual civilian clothes, something she wasn't sure she had ever seen him wear. His hair had grown out a little, and the scar on the left side of his face seemed to have faded somewhat. He was leaner, too. But it was definitely him, and he still had that air of command.

"Captain Lee," he said. "It's good to see you. I'm glad you survived the fight."

Lee said nothing. She was speechless, and she didn't want to say a something that sounded as awkward as she felt. Chambers was her uncle, something he had never revealed to her; she had only found out after he had gone missing after the battle of Alpha 51.

He had been a martial arts student of her father's, and he had eloped with her father's younger sister before Lee was born.

Growing up, Lee was aware, from whispers among the adults in the family, of the scandal that resulted when her aunt had run off with a foreigner. The matter was still fresh in everyone's mind when her aunt died of cancer far away from her family. Lee did not know at the time she served under him that Chambers had been that that young man who had eloped with her aunt Mei Ling, for whom she was named. She also did not know that her aunt had given birth to a son, Jay Cambers Junior, who had later become an Alliance Naval officer.

She felt a mixture of emotions. He was a good leader and she had always had confidence in his integrity, judgment, and leadership. But after the battle of Alpha 51, he had left her alone to face the wrath of the Navy: arrest, court martial, a kidnapping attempt. Raymond had claimed

Chambers had reached out to the Marines and the Chinese Consulate to help her, but Chambers hadn't been there, and the pain of that abandonment was still fresh.

Finally, Lee spoke. "Sir, thank you for making contact. We were hoping the arena fight would attract your attention. Our mission is, in part, to contact you."

Lee turned to Zakany, who had locked eyes with Nemeth. Lee said, "I think Laura and Matthias need some privacy." She waved to the now-empty bedroom. Laura nodded and went wordlessly into the room, followed by Nemeth.

There was a brief knock at the door, and Wasp came in. He scanned the room, saw Chambers, straightened up and said, "Admiral, it's good to see you again."

Admiral Chambers said, "Congrats on your promotion, Lieutenant."

Raised voices came from the bedroom. Laura and Nemeth were speaking in their native Hungarian. Everyone pretended not to hear.

Chambers turned back to the others in the room. "Johnson," he said. "Is it Sergeant now?" When she nodded, he said, "Congrats to you. I'm glad we got you off that rock during the battle of Alpha 51."

He turned to the Wells. "Sergeant Wallace and Corporal Jones. That was quite a performance you gave as a couple."

"Sir," said Wallace. "Can I ask how you knew who we were?"

"Don't beat yourselves up," said Chambers. "We check out all newcomers. You got our attention because you are more physically fit than the typical émigré to Ephesus. The provocation gave you away. After that it didn't take long to find out who you are."

Jones shook his head and looked ruefully at Wallace. She had been right, he thought. The fight had been a setup to see how he would react. He had blown his first job as an undercover operative.

One of the Marines who was working on a computer at a small kitchen table said, "Ma'am, I have Raymond and Lieutenant Danner in queue. We can bring them live on split screen whenever you say."

"Ask them to stand by," said Lee.

A moment later Zakany and Nemeth came out of the bedroom, both stone faced.

"Let's get started," said Lee.

CHAPTER 77

GOOD LUCK WITH THAT

Under Sergeant Johnson's supervision, the Marines arranged the various chairs in the room so that they were formed in a semicircle facing the computer screen, which was set on a low table. Chambers sat in the center, with Lee and Wasp on either side. The screen was split with Danner on the right side and Raymond on the left.

All eyes looked to Lee, who let a moment pass before she said, "We need to do a few things at this meeting. First, we need to establish what you, Admiral Chambers, are doing on this planet and what your intentions are." She paused and looked at Chambers to see if he wanted to comment. He looked back at her steadily and said nothing. Chambers, she knew, was unlikely to answer a question that had not been asked.

"Sir," she continued. "I don't mean to be rude, but we are under orders to determine your status and intent. You are technically absent without leave and, apparently, running a criminal enterprise. I do have the authority to take you into custody, should your answers prove unsatisfactory."

"Good luck with that," said Nemeth.

Lee ignored the implied threat, looked at Chambers and said, "Sir, I have some questions to ask you. I hope you will cooperate."

"Please continue," said Chambers.

"First, are you operating under the authority of any agency of the Alliance?"

Chamber shook his head and said, "No, and I would think you would know that."

Lee looked at Danner on the screen and he nodded.

Lee said, "That's good to know, sir. Lieutenant Danner originally got the mission to contact you from the Chief of Naval Intelligence, Admiral Bluefield. She told him she didn't think you were being run by one of the intel agencies, but she couldn't be sure because the agencies do not cooperate under the current chaotic circumstances."

"I knew Stephanie Bluefield many years ago," said Chambers. "She used to run special ops teams when she was a lieutenant. She is good, and I trust her judgment."

He paused and then continued. "Can I ask this: if Danner got the original mission, why isn't he running this meeting?"

Lee said, "Lieutenant Danner was severely wounded in a recent operation. His injuries have not healed, and he requires treatment at a level-four medical center, which is not available. He has refused evacuation and has ceded command to me."

"Understood," said Chambers. He turned his attention to the screen that held Danner's image and said, "Lieutenant Danner, congratulations on your promotion. I hope you heal quickly from your wounds."

"Thank you, sir," said Danner somewhat stiffly.

"And Chief Raymond," said Chambers turning his attention to the left side of the screen. "Good to see you again. I hope you have good intel on the mining operation. We haven't been able to get much in the past two years."

"Good to see you too, sir," said Raymond. "I'm happy to brief everyone once Captain Lee asks for it."

Chambers turned to Lee and said, "Ask your questions, Captain."

"Sir, tell us why you are here, what you have done while here, and what your intentions are."

Chambers started, "We came here based on intel from Chief Raymond that the mining of Boron Nitride in the system's Oort Cloud was potentially linked to separatist or terrorist activity. I felt the best way to determine the reason for that activity was to take over the criminal organization that we believed was running it. We were able to take control of the planet. But unfortunately, it turns out that the mining operation is compartmentalized under different leadership. Our planet-side organization does provide some support for the mining operation, but we haven't been able to get in to take a look, nor have any of our attempts to get someone inside been successful."

"And your intent going forward?" asked Lee.

"Originally, once we found out what the mining operation was about," said Chambers, "I intended to stop what they were doing if I had the resources or notify the Alliance if I did not."

The group was silent as they digested this, then Lee asked, "Was this operation being run by separatists?"

"No," said Chambers. "Certainly not. The operation we took over was a third-rate Mafia-style operation. We took control in one day. Someone put them up to this, but they didn't really understand it. They were running girls, gambling, and drugs, not espionage."

"So, where are you in your plan?" asked Lee.

"We're stuck," said Chambers. "We have control of the city, the space ports, and all the logistics. I've got almost nothing on the mining operation. It's good you came when you did."

"Admiral Chambers," said Danner from the screen. "One of the options Admiral Bluefield considered was coming in heavy with the Navy and taking over the mining operation. Is that an option you would consider?"

"First, it is not for me to consider or approve any options. Captain Lee has made clear she is in charge," said Chambers. "If you are asking my opinion, I don't think coming in heavy is a good option just now. A Naval battle group cannot be hidden. Once it's in the system, whoever is running the mining operation will destroy the operation and flee. We need to do more than stop what they're doing. We need to find out what's going on."

"I think Chief Raymond had some ideas about that," said Lee. Everyone had taken to addressing Raymond as Chief, even though he was no longer in the Navy. He welcomed this, or at least did not object.

All eyes turned toward Raymond on the screen.

"Chief," said Lee. "We want you to bring us up to date on your reconnaissance of the mining operation. But first, can you inform the Admiral about your theories on what the Boron Nitride is being used for?"

Raymond quickly and concisely explained to the Admiral his progress in achieving time travel and his belief that the separatists were doing something along the same lines.

Chambers, for his part, did not show any surprise at the revelation that time travel to a parallel universe was something Raymond had done repeatedly.

Raymond said, "Sir, you don't seem surprised by any of this."

"I'm surprised that time travel, or something like it, is possible," said Chambers. "I'm not surprised you figured it out. And you are perhaps the only scientist whom I would believe had accomplished this."

Chambers thought for a moment and then said, "What about the grandfather paradox? Doesn't time travel create the potential for a contradiction in cause and effect?"

"No, sir," said Raymond. "It would create a paradox if we could go back in time and change our own timeline, for example by killing an ancestor before he or she could conceive a member of our own line. But since we don't travel back in our own timeline, there is no paradox."

"That is an unsatisfactory answer, Chief," said the Admiral. "I'll need a more complete explanation later. For now, let's not hold you up."

CHAPTER 78

RAYMOND'S RECON

"Chief, please tell us about the mining operation. But before you start," said Chambers. "How did you get close enough to scope out the mining facilities? We've tried a dozen times, even using probes, and every time we get acquired and waved off before we can get close enough."

"Nanites," said Raymond. "Technically, they are nano-robots or nanobots that are specifically designed for reconnaissance. My firm has made them no larger than a particle of dust, so they are practically invisible to any of the scans used by most defensive systems. We can shoot them out at near light speed, they infiltrate a target, and we use them to report back in a variety of spectrum."

"And you used them to reconnoiter the Oort Cloud mining operation?" asked Chambers.

"Not only that," said Raymond, "But my nanites have penetrated all of the mining operations stations, including the mining vessels."

"So, you can see inside the facilities?" asked Chambers.

"Yes and no," replied Raymond. "I can see inside the facilities and vessels, but the nanites can't be everywhere at once. No single nanite can send an image by itself. They need to be together at a certain level, a critical mass, before they can transmit anything coherent. So, I need to move them around based on what information is needed."

"What are their limitations?" asked Lee.

"Good question," said Raymond, once again beaming as if at a bright student. "The nanites are not immortal. They burn out or sometimes just get lost. And they aren't intelligent. They can't improvise. I have to tell them exactly what to do, where to go, where to look, and what information to transmit."

"Can they be detected?" asked Danner from his image on the split screen.

"Yes," said Raymond. "Anyone who knows to look for them will detect them if they are close enough."

"Can they be traced to the source," asked Chambers.

"Hmmm...," said Raymond. "That is a good question. I'd say yes, but not immediately and not to the local source. This is an emerging technology. My firm is probably the leader in the field. It would not be so much of a leap to suspect they came from my research."

"Can they be used for sabotage?" asked Wasp.

Raymond was thoughtful for a moment and then said, "They can be programmed to interfere with electronics onboard a vessel or platform. If that is sabotage, then yes. But the limitation is, again, their lack of task flexibility. We would have to know exactly what we were asking them to do."

Wasp asked, "Can they be tasked to interfere with sensors?"

"Now that, they can certainly do," said Raymond.

Lee said, "Can you tell us about the results of your surveillance?"

"It is a fairly straight-forward set up," said Raymond. The group saw schematics pop up on the screen. "There are several major functions: command and control, maintenance, mining, and storage." There is a central base located on one of the larger asteroids. It serves as a headquarters and a maintenance depot. There are four mining vessels, each outfitted for asteroid mining. They go out individually to find deposits and mine them. There is a storage vessel, the Octavia, where the mined deposits are brought and stored. All of these are in a stationary orbit relative to the Oort Cloud itself.

Lee said, "Can you send us the schematic and other raw data?"

"Yes," said Raymond. "Doing that now."

Nemeth spoke up, "John, we noted the power output, can you tell us what that means."

"Well, I can tell you what I know," said Raymond. "Their power output at the storage facility is truly enormous. And although it fluctuates somewhat, in a cyclical pattern, it is continuous. And that's a problem."

"Why a problem?" asked Nemeth.

"It's a problem because I can't figure out why they would do that," replied Raymond. "We don't use continuous power when we hop to parallel universes. Instead, we do it in targeted bursts."

"Interpretation?" asked Lee.

"Almost certainly testing the time travel capability," said Raymond. "The first boost of power is needed for the determination of the timeline. The second is pushing something through or retrieving it. We use the same cyclical power output when we jump to other timelines, but nothing like this scale of power output or duration. I hate to admit it, but I think they are more advanced than we are."

The group was quiet for a long moment while each considered the significance of that.

"OK," said Lee. "We need to get organized for the next phase. First, let's recap where we are now."

Lee looked at Chambers and said, "I think we can agree that Admiral Chambers and Master Warrant Officer Nemeth are not working outside the interest of the Alliance. As such, sir you will not be arrested, at least not by me and not at this time."

"Glad to hear it," said Chambers with a wry smile.

"Am I correct in thinking, sir, that you control the ground-side spaceport facilities as well as the orbital maintenance platforms?"

"That is correct," replied Chambers.

"Do you have any combat power that could be brought to bear if needed?" asked Lee.

"We have a few outdated vessels, including one older D-Class frigate. That one is armed with missiles and cannons," said Chambers. "We use it for orbital defense. It's mostly for show to dissuade pirates from operating in the commerce lanes. It won't stand up to anything modern or well trained. In a pinch, however, it might be used."

"Good to know sir," said Lee. "I'll keep that in mind."

Lee turned to Raymond and said, "Chief Raymond, I assume you don't have an unlimited supply of nanites. What is your lead time to make more?"

"I can create a full batch every 48 hours," said Raymond. "That's the easy part. If they are going to be used against the mining facilities, I will need to program them for specific tasks. That could take longer depending on the complexity of the task they will be given."

"Understood," said Lee. "Here's what we'll do: Chief Raymond, I want you and Lieutenant Danner to develop an estimate on what the mining operators are doing, and what are the threats to the Alliance from those activities."

Lee looked at Wasp. "Lieutenant Wasp, using the information you have already received from Chief Raymond, and the information you will receive based on the analysis from Lieutenant Danner and the Chief, I want your team to work up operational options for various scenarios. I want you to include, at a minimum, options for further reconnaissance, selective intervention, capture, and destruction of those mining facilities. For any other possible options, I leave that to your judgment and imagination."

"Yes, ma'am," said Wasp.

"Let's meet back here in 48 hours and go through the options. Be prepared to brief those to me. I will decide at that time."

Everyone seems to acknowledge these instructions by nodding and getting ready to move.

"All right then, dismissed," said Lee.

CHAPTER 79

SECURITY

Two days later, the team assembled in a different safe house. Wasp was wary of staying in one place too long. The new safe house was a farmhouse well outside the city. Wasp preferred this location because it had good fields of observation; an adversary could not approach the house easily without being detected.

It was also more convenient for another reason: it had more room. There were a dozen bedrooms, and space in the barn for Wasp and his team to conduct their planning and rehearsals. In the house itself, there was a large central room that had been turned into an operations center. It had tables and charts placed on easels throughout. Danner and Raymond had both returned and been holed up with Wasp and his team.

As the time for the options briefing approached, the room was converted into a briefing room, with two computer screens at one end and chairs placed in a semicircle facing the screens.

A half hour before it was to start, a nondescript sedan pulled up with Nemeth driving and Chambers in the front passenger seat. Chambers came into the building and met Lee in the kitchen where she offered them coffee, which he accepted.

Nemeth met with Sergeant Johnson and took a walk around the perimeter to verify the security for the admiral and the meeting.

She showed him the various avenues of approach and the measures taken to cover them.

"You have a sniper in the house, I assume," said Nemeth.

"Actually, no," said Johnson. "Too obvious. Anyone attacking the farmhouse will assume there's a sniper there and will pound the windows to suppress the fire. Our snipers are in the wood line, there," she pointed, "and there," she shifted her hand to show him.

"And if there is an armored vehicle attack from the road?" asked Nemeth.

"We've got the road mined, and we have a concealed anti-tank grenade launcher offset there," she indicated a small shed set off from the barn.

"And if the attack is from the air?" he asked.

"Every one of our positions has fire-and-forget shoulder-fired anti-aircraft missiles. The seeker heads don't need to be guided by the shooter. Once fired, they will use a combination of infrared, radar and visual to track and attack an aircraft up to an altitude of 10,000 meters. Plus, we could call the shuttles from orbit into the defense, if needed."

Nemeth nodded and said, "Very good, Sergeant. You obviously don't need any help from an old spec ops guy. Thank you for showing me."

"Actually, I could use your advice," she said.

"Yes?" he asked.

"This entire defense assumes any attack will come from the road. What if it comes from the forest behind us? I don't have enough people to cover both."

"Good point," said Nemeth. "Do you have any remote sensors?"

"Yes, all Mark 2. They are not state-of-the-art, but they will do in a pinch," she said.

"Oh, the times I wish I had had a Mark 2 sensor," he said with a smile. "You could deploy those about 100 meters into the tree line. If anything comes that way, you will have time to redeploy your forces to meet that threat."

"Thanks, Master Chief," she said. "I think that will work." Then she brought her wrist to her mouth and began to give orders into her mic to emplace the remote sensors.

When she finished, she said, "So you were with the teams back in the day?" Teams was a euphemism for the Oracle operational detachments.

"Hah," he said with a smile. "You know we are not supposed to speak of such things." Johnson's face reddened in embarrassment.

"Please don't worry," he said smiling. "It is so long ago; nobody remembers or cares. Yes, I was on the teams many years ago. I was the Master Chief for Delta Squadron. Commander Chambers was the CO. We had many good missions then. Some good times and some sad times as well."

"I heard you were a pilot," she said.

"Yes, back in the day we had to qualify as pilots to be on the teams. Not a full qualification, you see. Just in case of an emergency, so we could fly what was available. When Commander Chambers had to leave the force from his injuries, I went to flight school and got my wings and my warrant. Then I flew for special ops exclusively for many years."

Nemeth looked at his watch and said, "We should be getting back; the briefing will start soon."

CHAPTER 80

MISSION BRIEF

Lee stood behind the podium and the computer screen and looked at the assembled group sitting in a semicircle facing her. In the center, next to her empty chair, was Chambers with Nemeth to his right, then Raymond. On his left were Wasp, Johnson and Zakany.

She began without prelude: "We're here to discuss options for dealing with the mining operation. Dr. Raymond has done additional research and together with Lieutenant Danner they have developed possible courses of action along with risks and payoffs. Once we've heard that, Lieutenant Wasp will brief on what capabilities could be brought to bear on the various options. Then we will discuss those options, and I will select the appropriate course of action."

She looked at Raymond and said, "Chief, you're on," and returned to her chair next to Chambers.

Raymond went to the podium, standing behind it. Always the professor, thought Lee.

"As we discussed two days ago," he began. "The mining operation has three major functions: command and control, mining, and what we

305

thought was storage and refining. In the last few days, we have been using the nanites to gather more information. We now believe the third function is not storage, but rather an active timeline acquisition and targeting center."

"Meaning what?" asked Lee.

"Meaning they are using the Boron Nitride to facilitate observation and travel to alternate timelines," said Raymond.

"Their purpose?" asked Chambers.

"I don't know," said Raymond.

"What makes you think it's not just storage?" asked Lee.

"Two things," said Raymond. "First, the power output is enormous and continuous, indicating they are performing what we call pathway jumps."

"And the second?" asked Lee.

"The sheer quantity of Boron Nitride going into the facility," said Raymond. "They've been shipping raw ore to that location for over two years. The so-called 'storage' vessel, the Octavia, is simply a converted cargo vessel. It would have been filled past capacity many times over and long ago."

"So, are they using it up somehow?" asked Lee.

"No," replied Raymond. "They are transporting it to an alternate timeline."

"For what purpose?" asked Lee.

"To store it for use in a later timeline," said Danner.

All eyes turned to the lieutenant as he stood and walked over to the front next to Raymond.

"Please explain," said Lee.

"It's the most likely explanation," said Danner. "They know they can't ship it out of the system in our timeline without risking detection. That much vessel traffic would attract attention. We think they are moving back in time to a point before FTL was developed, then moving it to a location where they will 'find it' in our current timeline, and then use it as they see fit to accomplish their broader aims."

"How would this be done?" asked Lee.

"Remember when Chief Raymond wanted to prove to us that he could go back in time? He did that by going back in time and burying an artifact near the monastery and then returning to the current timeline and digging it up."

"You think that's what they are doing?" asked Lee.

"I'm sure of it," replied Danner.

"Where are they taking it, and when are they taking it?" asked Lee.

"We are working on some screening and evaluative criteria," said Danner. "But the best way to find out is to take that facility intact."

"I agree," said Lee. "What are our options?"

"Of the three functions," said Raymond, "the highest payoff is the Octavia, what we consider the storage vessel. Gaining unrestricted access to that would allow us to discover where and when they are sending the Boron Nitride. If it's not possible to take the facility intact, then the next best option is to destroy it. That's not ideal. We would not be able to determine where and when they are taking the Boron Nitride, and they could just start up from another location and continue."

"What about the command-and-control center?" asked Chambers. "Why not take that? Surely that's where the planning is taking place."

Danner answered, "No doubt, sir, that is important. But again, it is a secondary objective." He looked at Wasp and said, "I think it will come down to what capability the special operations detachment can bring to bear. I'm guessing here, but I doubt 12 operators could take both facilities."

All eyes looked to Wasp. He stood and said, "We are a little ahead of ourselves just now. We do have options for every course of action. But since it has been raised, Lieutenant Danner is correct. We can take the storage or the command-and-control facility, not both simultaneously."

"How would this be done?" asked Lee.

"With help from Chief Raymon's Nanites, we can hijack one or more of the mining vessels bringing refined ore to the timeline facility. Dock at the facility and then take it over."

"What about their defensive response?" asked Chambers. "Every time we tried to get close with probes, they launched missiles to intercept them."

"Sir," said Wasp. "We have a plan for that, but as you will know, that option has uncertainties and risks. Chief Raymond believes he can use the nanites to confuse any defensive response the miners might bring, but the level of risk is high because it is an unknown element. We've never used nanites before."

Just then they all heard the unmistakable boom, boom, boom of spacecraft entering the atmosphere. Wasp and Johnson stood and immediately began listening to their communicators.

Lee asked Wasp, "Are those our shuttles coming in?"

"No ma'am," replied Wasp calmly. "We are under attack."

CHAPTER 81

A FIGHT AT THE FARMHOUSE

"Patch me through to your tactical net," said Lee. "And put up the tactical situation on the screen here."

Lee turned to the computer screen and saw schematics showing inbound aircraft moving at supersonic speed, still trailing ice particles from the upper atmosphere. The schematic showed the defensive positions in and around the farmhouse and outbuildings.

Nemeth got up and moved to Johnson and said, "Sergeant, I'm with you. Please put me to use. I have a sidearm; do you have anything larger?"

"You bet, Chief," she responded. "Let's get you a sniper rifle." They both left the room at a quick but controlled pace. Both had been in combat many times. It was not a time to show panic.

Zakany headed for the door and said, "I'll set up a med station in the barn. Lieutenant Wasp, please send me one of your medical corpsmen." And then she was gone.

Chambers was on his communicator speaking urgently and then he signed off and turned to Lee and Wasp. "It's not my people. They don't have spacecraft and they don't know anything about this."

"Can you call in assistance?" asked Lee.

Chambers shook his head. "They're not much use in a fight like this. We've been intentionally declawing them over the past two years. No heavy weapons, just sidearms, and they can't shoot very well." He shrugged. "Best I can do is get surveillance on any movement our way coming from the town. I'll have them put up barriers that will slow down any vehicle movement. If the enemy comes this way from town, and then needs to dismount to clear the road, my guys will take shots at them. But I have ordered them not to close with the enemy, long distance shots only."

Lee walked to one of the open windows. She heard the whoosh and saw the distinctive plume of the simultaneous launch of shoulder-fired anti-aircraft missiles.

"Any ground threat yet?" asked Lee.

Just then they felt more than heard an explosion that rocked the house. The lights dimmed and dust settled down from the ceiling.

"That's mortar fire," said Wasp. "It is originating six kilometers out coming from the forest side."

"Do we have counter-battery capability?" asked Lee.

"Yes," said Wasp. "I'm bringing the shuttles back, and I'll try to get an air strike on that mortar."

Wasp spoke again into his communicator, then turned back to Lee. "Ma'am, there's a vehicle assault from the road and a dismounted attack from the wooded area."

"Do you have enough resources to handle both ground assaults and the air attack?"

Wasp paused for a moment and then said simply, "No, ma'am."

"Understand," said Lee. "Let's evacuate. We have no reason to defend this farm."

"Roger, ma'am," said Wasp. He turned to his communicator and said, "Execute phased withdrawal Alpha, and move to designated pick-up points."

CHAPTER 82

A FIGHT IN THE WOODS

When Nemeth heard over his coms that an attack was coming from the woods, he had asked Johnson to let him take that direction until she could send reinforcements that way. He was comfortable in the woods, having spent much of his youth in Hungary hunting in the heavily forested area outside Budapest. He had also trained and fought in jungles in his many years in special operations.

On his handheld he could see a schematic of what the mark 2 sensors were providing. It was a squad of eight, and they were moving quickly toward the farm using a deer path that led to a meadow adjoining the farm.

Nemeth spoke into his coms. "Sergeant Johnson, I'm in the woods. I'm looking at a squad. I can handle this. I know you need to evacuate. No need to send anyone this way. Good luck."

"You're not coming with us?" She asked.

"No. This is our turf. The admiral and I will stay put," said Nemeth. "These people are after you, not us. Once you are gone, they will leave here and lick their wounds."

"I'm glad I gave you that sniper rifle," she said.

"Yes, thank you," said Nemeth. "I'll get it back to you when I can. But I won't be using it."

"No?"

"No," said Nemeth. "This is close-in work. I'll use my knife and sidearm. Gotta go."

Nemeth moved to a concealed spot near where the squad would need to pass on their way to the farm property. His pulse and breathing lowered and he readied himself for the battle to come. If he were honest with himself, he would acknowledge that he loved battle, and especially close-in fighting.

As they came closer, he could tell they had some military training, but they were not true professionals. They were bunched too close together and looking directly ahead instead of side to side.

He let the first six of the squad of eight pass his position, then he stood up into a firing position and used his sidearm to fire rapidly, one shot per adversary, each one a kill-shot to the back of the head. Three down before they could even react. Then the remaining five did exactly the wrong thing, just as Nemeth knew they would. Instead of diving off the path and out of the line of fire, they all turned to look in the direction of the firing.

For three of them, it was their last mistake. As they turned, Nemeth continued firing — four, five six — all down with perfectly placed deadly head shots.

The final two were actually behind him. Nemeth dove into the vegetation and rolled away. As expected, the two still-living soldiers fired wildly into the brush.

In a moment, Nemeth was up and running directly toward the two. He burst out of the vegetation within two feet of the lead soldier. He ran past him, slashing with his knife horizontally across the man's throat, a killing blow. The last man looked at Nemeth with wide, terrified eyes. He tried to raise his rifle, but Nemeth was faster. He closed the distance, slammed his fist into the man's throat, and snatched the weapon out of his hands.

The man was on his back, holding his hands to his throat, gasping for air. Nemeth pointed the man's rifle at his chest and said calmly, "Surrender or die." The man, still unable to speak from the blow to his throat, nodded his head and raised both his hands in the universal gesture of submission.

On his coms, Nemeth heard Johnson say, "Chief, we're out of here. If you need help, we could swing by and lend a hand." He could hear the engines of the three shuttles lifting off near the farmhouse.

"Not necessary, Sergeant," he replied. "Seven down, one in hand. We'll see what he has to say. How about at your end?"

"We damaged two of the three attacking aircraft. All three of their drop ships retreated," she replied. "We got everyone out but the Admiral, who is staying back. No fatal casualties on our team, though we do have some scratches. Dr. Zakany is with us and caring for the wounded. We did leave a mess; sorry for that. There are several wrecked vehicles, a dozen bodies here and there, and the farmhouse is on fire."

"Not to worry, and well done, Sergeant," he replied. "I'm glad to see Oracle still has the touch."

Nemeth turned back to his captive, and said to the terrified man, "Now let's find out if you can be useful to us."

CHAPTER 83

LET'S DO THAT

Lee sat in the ready room of the command shuttle. Wasp sat beside her, with Raymond, Johnson, Danner and Zakany connected by video from the other two shuttles. Chambers and Nemeth were connected from Ephesus.

Lee said, "First, let's do a quick update on the battle of the farmhouse. Thoughts on who attacked us, why, and how they knew where we were." She looked at Wasp and said, "Lieutenant, please go first."

Wasp said, "The attacking force was not native to the planet. The three drop ships jumped in from FTL into low orbit and immediately descended to the farmhouse. The forces for the ground assault were pre-positioned and timed to coincide with the air attack. We had three vehicles attack from the road, and a dismounted squad movement from the forest side."

"Sergeant Johnson," said Lee. "You were responsible for perimeter defense. Please tell what happened."

"Ma'am," said Johnson. "The first attack came from the drop ships entering the atmosphere. It was clear they intended to attack, and I gave the antiaircraft teams the weapons free command. All three of them launched shoulder-fired surface-to-air missiles. Two of them found their targets and damaged their aircraft. The third was intercepted by counter-missile fire from the drop ship. All three drop ships departed the area without firing."

"And the ground assault?" asked Lee.

"The three vehicles attacking from the road were standard civilian pickup-type vehicles that had been augmented with reinforced armor and had a mounted 20-millimeter cannon on the cab in back," said Johnson. "The admiral had watchers in place who engaged the vehicles from a distance. Those watchers had also placed obstacles in the road. That slowed the attackers down a few minutes and allowed us to better position ourselves to engage those vehicles once they got closer."

"You had some minor casualties," said Lee. "What was the source of those?"

"Two sources. First the pickups did get off some effective fire before we destroyed them. Also, we took light mortar fire from a position in the forest. All injuries were minor and have been treated by Dr. Zakany and our corpsman."

"And the attack from the forest side?" asked Lee.

"Master Chief Warrant Officer Nemeth handled that," said Johnson.

Lee looked on screen at Nemeth, and asked, "Chief, can you tell us what happened at your end?"

"Sergeant Johnson had deployed mark 2 sensors on the forest side," said Nemeth. "Once we knew there was an attack from that direction, I got her permission to investigate. What I saw from the sensors, and later confirmed by sight, was a squad of eight moving down a path toward the farmhouse. As they were clearly not professionals, I asked permission to handle them myself so that the rest of the team could concentrate on evacuation."

"And you handled them yourself," said Lee. It was a statement, not a question.

"Yes, but this was not difficult," said Nemeth. "As I said, they were not fully competent. They were bunched together and not looking around, but only in their direction of movement. I used my sidearm, then my knife."

"And you have a prisoner. I understand he was the only survivor of the ground force," said Lee.

"Yes," said Nemeth. "He says he was hired as an independent contractor on earth. After a few weeks of training, this was the first time he was sent on a combat mission. He doesn't know who was really paying him, and he wasn't told why his squad was supposed to attack the farmhouse."

"Is he telling the truth?" asked Lee.

"Yes," said Nemeth. "These are not professionals. He may know more than he thinks he does, and I will continue to gather facts about his recruitment and training. Perhaps that will give us a better idea of who is behind the attack."

"So now let's talk about why they attacked us and how they knew where we were," said Lee. "Lieutenant Danner, you are the intelligence officer on this mission. What have you got for us?"

"These were clearly relatively untrained mercenaries that some unknown party hired on short notice to try and remove or impair our ability to conduct further activities," said Danner. "They weren't told they were going against a special ops force, and they clearly weren't prepared for what they found."

"So, what does that tell us?" said Lee.

Danner said, "I think we have to assume that whoever hired them knows we pose a threat, but it is unlikely they knew we were special ops."

"If I may," said Chambers.

"Please go ahead, Admiral," said Lee.

"First, let me apologize that an attack of any kind took place in an area that I controlled," said Chambers. "These mercenaries could not have come in undetected by my people, and they didn't. They had the cooperation of people who should be loyal to me. Apparently, I have my detractors since I closed down prostitution and drugs. They gambled that I would be killed in the attack. Alas, a miscalculation."

"So, you have been questioning your people?" asked Lee. "What have you found out?"

"Only that whoever paid off my people to look the other way, had a lot of money. They used a third-party to broker the deal, so there's no trail to follow, at least in the short run."

Lee paused a moment to gather her thoughts and then said, "I'm thinking this attack was ordered by whoever is running the mining operation. And that means our element of surprise is blown. Any disagreement?"

"Yes and no," said Danner. "Yes, the originators of this attack are probably also behind the mining operation. They detected us and moved against us on the general principle that any unknown force would be a potential threat. But, no, it is very unlikely they know of our plans to interfere with or capture any portion of the mining operation. As far as they are concerned, their operation is safe operating out in the Oort Cloud. They have their defenses and can't possibly know about Dr. Raymond's nanite scouts."

"Have we shown our hand by the very capable defense put up by Sergeant Johnson and her team?" asked Lee.

Danner replied, "Yes, but it may take a while for that realization to filter back to whomever is behind this. They didn't know in advance that this was a highly trained special operations force. If they had, they wouldn't have thrown an untrained group of wannabe mercenaries at the farmhouse. All they will know right now is that the attack was a failure. They won't necessarily know why."

"So, what you're saying is that we still have some window of opportunity to act against the mining operations."

Danner said, "Yes, ma'am."

"Any other thoughts?" asked Lee. "Lieutenant Wasp, your people will need to conduct any action. Are you willing to go forward with the risk that they might expect you?"

"Ma'am, I agree with Lieutenant Danner," said Wasp. "We are in route to the mining operation now, and we can be there in two hours using our FTL. There is very little chance the mining operation could digest what's happened at the farmhouse or anticipate the threat we pose. Now is the time to act with what we have. If we wait, they will eventually take precautions. Then we'll need a much larger force, and that won't be ideal."

"Sounds good," said Lee.

Turning to Raymond, she said. "Chief, can your nanites be ready in two hours?"

"They are ready now," said Raymond.

"Lieutenant Wasp," said Lee. "Can your force be ready in that time?"

"Yes, ma'am," said Wasp. "We've been rehearsing at the barn these past two days in case we were needed on short notice."

"All right then," said Lee. "Let's do that."

CHAPTER 84

WAITING FOR PREY

Sergeant Johnson, with three of her Marines, were in full battle armor, the Extra Vehicular Mobility Unit Suits or EVMU, the so-called moo suits, designed for operations in vacuum. They were sitting in a rough circle on an asteroid in the Oort Cloud of the system. There was almost no light this far from the system's star which, because of its great distance, looked only slightly brighter than the other stars in the sky.

They had been deposited here only an hour ago, and patiently waited for their prey. During the battle of Alpha 51, she had been stuck on an asteroid because her booster failed to fire. That had been a misery for her. Not only had she had to stay there for hours waiting to be picked up by the Marine Search and Rescue team, but worse, she had missed out on the assault on the moon base. One of her team had been killed during a rough landing on the moon. Although she knew it was unreasonable, she felt guilty at not being there. Maybe if she had been present, she could have made a difference.

Now she was back on an asteroid, light years from any planet, on an operation that had been thrown together on the run. This time she wasn't a private. She was a sergeant, the deputy commander of an Oracle team in combat. She felt the weight of the responsibility for the lives of others.

She had served with Wasp for years, long before either of them had joined the special ops community. He had gone off to OCS, and she never expected to see him again. But then he had come to her at Paris Island where she was serving as a Marine drill instructor. With the advent of war, enlistments were up, and more instructors had been required. She liked it well enough, but she missed the thrill of combat.

Out of the blue, Wasp had shown up, told her she had a chance to serve with him again, with the best the Marines had to offer. That had been enough. He had always made being in charge look easy. He even had a sense of humor, at least when he was a gunnery sergeant and later a sergeant major. Less so now that he was an officer.

She saw it first on her heads-up display, the clear sign of a fusion engine firing to slow down. Their target was inbound, one of the mining vessels heading for this asteroid. Her team's mission was to take it.

Raymond had used his nanites and his intelligence resources to find out who the pilot was, and discovered the unfortunate circumstances of his employment. This information should be useful in gaining his compliance.

CHAPTER 85

STEALING A STARSHIP

Captain Raphael Cortez, the commanding officer of the mining vessel Phoebe, approached a large asteroid. It was the second stop made this outing. Soon its storage capacity for unrefined Boron Nitride would be at its maximum, and then it would need to travel back to the storage facility and unload its cargo. Then he would do this again.

When he took this job two years ago, Cortez was excited, and the promised pay was too good to ignore. He had been hired by a somewhat obscure company on earth that he had never heard of in his 20 years in the mining business. All they wanted, they said, was a captain for a six-month mining operation in the Oort Cloud of some out-of-the way star system about 100 light years from earth.

As the end of his six-month rotation neared, he asked about arrangements to go home. At first, his employers offered him more money to stay on, a lot more money. Another six months and he would be wealthy, able to retire in comfort and complete security.

As the end of his that second six-month tour approached, his inquiries were again met with offers of more money. But by then, he had

had enough. He wanted to go home, to see his wife again, his dog, his own house and bed.

But it was not to be. This time when he insisted on returning home, his employers sent him a video clip of his wife, bound, bloodied, and terrified. The message with the video said only, "We'd like you to stay on for another tour. Then you can go home."

So, he stayed on. He couldn't think of an option to get out of this. If he refused, surely, they would kill him and his wife. Maybe they would do that anyway. But what could he do? Choose certain death now or probable death at some future date? He chose to stay on.

The vessel was almost completely automated. He was, in fact, both the captain, and the entire crew; just him, and that was part of the problem. He was lonely. He missed his wife terribly; he regretted not loving her better when he was with her, and he felt intensely the absence of children in their marriage.

He maneuvered the Phoebe into position and used the automated claws to secure the asteroid. As per his operating procedure, he transmitted his status to the mining operations headquarters.

"Phoebe in position, all green here. Beginning ore extraction."

There was a slight delay in the response, which was unusual. A voice came across the speaker. "Roger, Phoebe. What is your estimated dwell time?"

Now that was different. In two years, no one from headquarters had ever asked that question. The dwell time — how much time he would spend on the asteroid — always depended on what ore they found and the Phoebe's capacity for storage.

He said into his communicator, "Say again, headquarters?"

The response came immediately. "Stand by, Phoebe. Prepare to be boarded."

Cortez felt the thump of something outside the hull of his vessel. He checked his sensors, but they showed nothing at all. How was that possible?"

For the first time in an exceptionally long time, he was terrified. Maybe this was his employers finally deciding they didn't need him anymore. Were they coming to kill him? But why do it out here? And why jinx his sensors? He could be killed anytime back at headquarters, where he would need to go anyway after dropping his load of ore.

His display now showed that his airlock was cycling, meaning someone was coming aboard. Panicking, he pushed the emergency disengagement switch on his console. This was designed to set off explosive charges that would thrust his vessel away in case the asteroid became unstable. Nothing happened.

Cortez took a breath and calmed himself. There was nothing he could do. He didn't have a weapon. And though he desperately wanted to go home, he had to admit he was tired. Dead tired. If this was the end, let it be quick; let it be over. He only hoped his wife would be spared.

He turned and faced the bulkhead door to the bridge and waited. The handle moved and the doors opened. To his surprise, a person wearing full body armor entered. This person had a rifle, but it was aimed downward, not at him. The figure removed its helmet and he saw a young woman, shaved head and fierce looking. He stood opened-mouthed, not knowing what to say. What could be said?

The woman smiled and said, "Hello, Captain Cortez. I'm Sergeant Johnson of the Alliance Marine Corps. I'm here to solve your problems."

CHAPTER 86

WE HAVE A PROBLEM

"That's the fourth mining ship," said Wasp over his coms.

Lee responded, "Roger that; well done."

Wasp and his team had done what they said they would do. With Raymond's nanites spoofing the mining vessels' surveillance system, his teams had taken all four mining vessels without a shot.

The vessels' pilots, it turned out, were basically prisoners being forced to fly for whomever was running the mining operation; as a result, they were all glad to have been 'rescued' by Alliance military personnel.

"Are the pilots willing to assist with the next phase of the operation?" asked Lee.

"For the most part, yes," Wasp replied. "All of them have family members being held as hostages back on earth. I have sent a message to Oracle teams there to go round them up. If those family members are still alive, we'll find them and bring them to safety. Once the pilots heard that their families would be safe, they were willing to cooperate. All we need them to do is dock with the Octavia, which is their normal procedure anyway. After that, it's all on us and they can sit it out."

"Walk me through the procedure. How do they normally offload the ore?" asked Lee.

"They tell us it is an automated procedure. The mining vessels dock at the specified portal. The pilots don't do anything. They remain onboard and the ore is taken off via an automated procedure."

"So, no interaction between the two crews?"

"Correct."

Lee said, "Chief Raymond, are we on track to jam the surveillance and coms for the mining command vessel?"

"It's better than jamming, Captain," said Raymond from his post in the Combat Information Center or CIC on the command Shuttle. "The nanites are creating false readings. The command vessels won't see they're being jammed at all."

"The storage facility is the real target," said Lee. "How're the nanites doing there?"

"They are fully integrated there as well," replied Raymond.

Lee wasn't comfortable. They hadn't had time to do a full rehearsal or to work through all the possible countermeasures the enemy might take. She knew you couldn't always have perfect intel going into an operation. The attack on the farmhouse had forced her to accelerate the timeline. Waiting for perfect information on the enemy would have invited disaster. Wasp was right. Now was the time. But that uneasiness remained.

For once, she wasn't in the thick of it, and she didn't like that. She was the commander of the entire operation, and thus she was back in the command shuttle directing activities. She had a lot of adrenaline and nowhere for it to go.

Worse still, they had been forced to violate one of the principles of war: to always keep a reserve. And she had no reserves. Everything they had was committed. If any of the team, now split in four parts, got into trouble, there was no clear plan to help them out.

"And they can't see us?" asked Lee.

"That is correct," said Raymond. "All of their scanning devices in every spectrum have to filter through a firewall my nanites have set up.'

Lee shook her head, and Raymond said, "Captain, you don't look satisfied."

"It's my job to be unsatisfied," snapped Lee.

Raymond looked steadily at her through the screen and said nothing.

"What about visual?" asked Lee.

"Everything, including visual light," said Raymond. "They cannot see us."

"I don't mean using cameras in the visual spectrum," said Lee. "I mean using their eyes, like through a viewing portal."

Raymond was silent, his eyes going wide with realization. "Shit," he said. "Let me check for viewing portals." A pause of a few seconds. Then. "The Octavia does have a viewing portal on its starboard side. It's not used for navigation, however. It's in the crew lounge."

"Do we have anything within their visual range?"

"Stand by... Nothing is there now, but two of our shuttles passed that way in the past hour. I can't be sure no one saw them."

"Show me the track on my screen," she said.

A second later she watched as her screen showed a schematic of a shuttle as it passed by the mining storage facility. Superimposed was a fan showing the range of what could be seen from the vessel's viewing portal. Sure enough, at a distance of about ten kilometers, the shuttle passed directly through the fan. She waited a moment and a second scenario appeared showing a second shuttle also passing within the viewing fan of the storage vessel.

"I am so sorry, Captain Lee," said Raymond. "That is entirely my fault."

"No, it's my responsibility," said Lee. "Let's deal with the problem we have, not the one we wish we had." She paused and then said, "How long was each shuttle within sight of the viewing portal?"

"The first one was 2.7 seconds; the second one was 4.3 seconds," Raymond said.

"Did the Octavia show any sign that it had seen an unauthorized vessel so close to it on either occasion?" asked Lee.

"What am I looking for?" asked Raymond.

"Think what they would do if they had a visual sighting of a craft that did not show up on their other sensors," said Lee. "What would you do?"

"I'd recheck my sensors, try another scan, look for other discrepancies," said Raymond.

"Did they do that?" asked Lee?

"Give me a moment," said Raymond. "Let me check their logs."

A minute later Raymond said, "There it is." She saw a list pop up on her screen. "One minute 40 seconds after the first shuttle passed the viewing portal, the pilot ran systems check and then, when all checks passed, they tried to run a radar sweep of the same area, and the sweep came back with nothing."

"So, they know their system is being spoofed," said Lee.

"Yes, they must," said Raymond.

"Lieutenant Danner," said Lee. "Are you listening to this?"

"Yes, Captain," said Danner.

"Please use that magnificent brain of yours and give me some options," said Lee. She knew this was Danner's true strong point. He could think things through front to back, back to front, look at all the options and pick the one that had the highest chance of success within the acceptable amount of risk.

Danner paused for an uncomfortably long time before answering. After a minute, Lee asked, "Andy, are you with us?"

"Yes, Captain," said Danner in an uncharacteristically formal tone. "Here are my thoughts: our objective is to take the storage facility intact. Normally, we would plan that so that the option selected would be within

328

an acceptable level of risk to our forces. If it couldn't be done within that constraint, we would choose another option or abort."

He paused again, and then said, "Because of the rush placed on us by the attack on the farmhouse, we have not done a full planning workup. We have never set an acceptable risk level, nor an abort criterion, nor selected other less desirable options."

"And where does that leave us?" asked Lee.

"I say we do that now. The last mining vessel is just now finishing up. We have about an hour before the first one arrives back at the storage facility. But we might not have that much time."

"Why not?" asked Lee.

"Chief Raymond," said Danner. "Has anyone in the mining operation sent out a distress call?"

"Certainly not," said Raymond. "My nanites would have prevented it."

"But has anyone tried to send a distress signal?" asked Danner.

"A moment," said Raymond as he checked his display and punch keys. "No, there hasn't been any attempt at a call for assistance."

"Danner, what's your point?" asked Lee sharply.

"They know there are hostile vessels in their vicinity; and if they know their sensors are being spoofed, why haven't they called for help?" asked Danner.

"Tell us, please," said Lee with ice in her voice.

"It's because they don't need to," said Danner. "They have a fail-safe procedure."

"Oh, shit," said Lee, suddenly understanding. "It's because the absence of a periodic message saying everything is OK, is the distress signal."

"Exactly," said Danner.

She went to her external coms: "Lieutenant Wasp, we have a problem."

CHAPTER 87

A NEW PLAN

"Go for Wasp," came the reply.

"We now believe the storage facility is aware of our presence and that they have likely sent a distress signal via a fail-safe procedure," said Lee.

"Yes, ma'am," he said with complete equanimity.

"Well..." Lee stumbled. She was expecting him to show alarm or be angry. "What can we do now? We may not have much time and we won't be able to take them by trick as we had hoped."

"Yes, ma'am," said Wasp. His flat voice was frustrating in the extreme.

"So, what's your recommendation?" said Lee.

"Oh, we have a plan for that," he said with complete confidence. "I thought you knew."

"No, I did not know," said Lee with anger. "When were you planning to tell me?"

"I have told you," Wasp said. "It's in my written assessment of options, delivered to your inbox..." he paused, "four hours and seven minutes ago."

"What the f..." said Lee. She took a moment and scrolled through her messages. "OK, I see it. Next time mention something like that to me."

"You were busy, ma'am," said Wasp. "And so have we been."

Both were silent for a long moment.

"I guess I'm waiting for your approval to proceed," said Wasp in that impossibly calm voice.

Lee quickly read through Wasp's proposal. She was still angry and a little embarrassed. But she had to admit that once again he had come through in a clutch.

"Approved," she said. "Let's do this.

CHAPTER 88

A HAPPY MARINE

Corporal Baxter Jones was happy. Finally, he was on an operation. After their cover had been blown on Ephesus following the cage fight, Wasp had ordered them to rejoin the team. Jones was so glad to be taken off of undercover duty. He knew it was part of the job, and that he would need to do better in the future, but his true love was combat operations.

The fight at the farmhouse had been satisfying, but brief. He had been in one of the defensive positions and had launched a shoulder-fired, anti-aircraft missile at one of the drop ships. His missile had hit home, and the damaged drop ship had been forced to retreat. He had also used a grenade launcher to take out one of the armored wheeled vehicles that attacked from the town. That was awesome, but brief.

Now he was pumped up and maybe just a little bit frightened. He was with a four-Marine team led by Sergeant Wallace. Man, she was cool under pressure. He had to admit, she was the consummate professional Marine leader. She had given him hell back in Ephesus for screwing up his undercover identity. At the time that had seemed unfair, but he realized later that she was right on. And then when it came time to report his failure to Wasp, she had taken responsibility. That was the Marine way. He would never forget that example.

They were in moo suits anchored to the outside of the hull of the storage facility, the Octavia. Each team had arrived on board one of the four mining vessels. The vessels had unloaded their cargo, just as always, via an automated procedure. While that was happening, the Oracle teams had departed the mining vessels via a rear airlock and now had taken positions preparing to make a breach.

Their task was to wait for the mining vessel to depart after unloading, then to blow the hatch with explosives, make a forced entry and capture the storage vessel intact. Once inside, the four teams would move to designated areas of the vessel to ensure its capitulation. This was what he was trained for. This was a delight to his soul. He had been given the job of being first through the hatch once it was blown open.

The explosive charge was set on a timer. Jones and all the team were listening to a countdown in their headsets. With five seconds remaining, he heard Wasp call out over the net, "Fire in the hole, Fire in the hole, Fire in the hole!" It was the traditional warning that an explosive charge was about to be detonated, and allowed any who were within the blast radius to get out of the way.

With one second remaining, Jones pushed off from his anchor to position himself to make the initial entry. He was technically in free fall, using only his stabilizers to maintain his position. The instant the blast blew out the hatch he would use his moo suit thrusters to enter the vessel.

And that's when everything went wrong.

CHAPTER 89

AND THAT'S WHEN EVERYTHING WENT WRONG

The instant the explosives blew the hatch, there was a tremendous flash of light, far brighter than what the explosive should have caused. Jones' three teammates, including Sergeant Wallace, were thrown clear of the vessel like rag dolls, speeding past him out into the darkness of space.

"Sergeant Wallace," said Jones into his coms. "Are you OK? Status?"

Nothing came back. He could just barely see the figures now, like plastic soldiers thrown by a child, their bodies spinning wildly.

"Any station this net, this is Jones," he broadcast. Nothing.

His brain struggled with what had just happened. What could have caused them to be flung away like that? Surely not the explosive charge. None of them were within the blast radius, and why wasn't Jones also flung away?

Then came a voice barely audible over static. "Jones, this is Johnson. What's your team's status?"

"Some sort of explosion, all three of my teammates were blown clear. I've got nothing back from them on coms."

"Same here," came Johnson's distorted voice. "Must have been an electrical charge meant to protect the storage vessel. Looks like anyone in contact with it got thrown clear."

"Your orders, Sergeant?"

"Stand by," came her reply.

"Captain Lee, this is Johnson, over."

"Go for Lee," came the response.

"The vessel was booby-trapped," she said. "It looks like the explosive charge triggered some sort of electromagnetic response. All but Jones and I were incapacitated and thrown away from the vessel."

"We see them on our scans. Moving the shuttles to pick them up now," said Lee. "Stay where you are, and we'll come pick you up as soon as we get the others."

"Negative, ma'am," said Johnson. "Jones and I will complete the mission."

"Sergeant, you've just lost ten members of your team, including your commanding officer. What was the abort criteria?"

"Ma'am, there is no abort criteria," said Johnson. "Lieutenant Wasp designated the capture of the Octavia as a primary concern."

The term primary concern was a rarely used designation that meant any and all efforts would be made to accomplish the mission. It was basically a do-or-die command. Even if only one remaining team member survived, that person would continue with the mission.

In all her time in the Navy and later in the Marine Corps, she had only known of its use one time before, and that was by Wasp when he conducted a hostage rescue of Chief Raymond. Now Wasp had given that order to his team for this mission, but he had left out any mention of it in his mission proposal to Lee.

Lee thought quickly about how to manage the situation. She could, of course, override Wasp's order. He was her subordinate, and he was out

335

of the picture, at least for now. But if they aborted the mission now, they would almost certainly lose their one chance to understand what the enemy was doing. She knew it was likely that some sort of fail-safe procedure would bring enemy forces to aid the mining operation. And she had three shuttles, that's all. If even one competent enemy frigate showed up, she would lose her entire force.

Finally, she said, "You will continue the mission, but I need you to stand by for some help."

CHAPTER 90

NO PLAN SURVIVES CONTACT WITH THE ENEMY

Lee had never done so much in so short a time. She knew that time was against them.

First, she had to direct her three shuttles to recover the members of Oracle that had just been blown away from the storage vessel. Retrieving them had been relatively easy, because the suits had automatic retrieval beacons that were triggered anytime the occupant was unconscious or dead. But it had taken precious minutes.

To her great relief, all of them were still alive. As she had thought, the storage vessel had given off something akin to an electric shock. It had rendered the occupants unconscious and fried most of the moo suit's electronic components. Luckily, the auto retrieval beacons were protected against just such a surge.

The wounded had been transferred to the sick bay aboard the command vessel where Dr. Zakany, aided by the two Navy corpsmen, had set up an emergency room to treat the arriving patients. They all showed symptoms of electric shock, some with burns on their torso where the suit's electronic equipment had been shorted out by the electric charge. Most were unconscious or semi-conscious.

While she was doing that, Lee called Raymond. "Chief, what was that blast and how come your nanites didn't prevent it?"

"My fault entirely," said Raymond. "The nanites only do what they're told. I had them totally focused on spoofing the vessel's surveillance system and hadn't considered that there might be a charged electric defense system."

"We're about to enter the vessel with a greatly reduced boarding party. Is there anything else we should know?"

"I'm sorry, I just don't know, nor do I have any way of finding out," said Raymond. "You've clearly lost all surprise. I think any automated defense systems will be activated. I just don't have anything else to offer."

"Do you have any nanites left?" asked Lee.

"I have a new batch brewing; it won't be ready for another two hours. But..."

"But what?" asked Lee sharply.

"I could program what I have now and send them out; their impact will be minimal."

"I'll take what I can get," said Lee. "Please do it."

Wasp was one of the first to regain consciousness, and he contacted Lee right away for an update. She came to the sickbay and told him what had happened and what she planned to do.

"I'll go with you," he said.

"You're in no shape to command," she said.

"You're right," he said right away. "You are the overall commander, that is clear. Johnson needs to stay in command of what's left of the Oracle boarding party. She's fully capable and there's no need to keep switching out commanders."

He looked at her directly and said, "But I am fit to go in. You know I wouldn't say it if it wasn't true. And I think you could use a hand."

Wasp quickly donned a new moo suit, and he joined Lee, who was already suited up and in the airlock.

Lee had left Danner in control of the CIC. His voice came over the coms channel. "Once you see the green light above the airlock, you will know that we are correctly positioned ten meters from the port-side docking bay. You can use the thrusters on your suits to cover the distance. The docking bay door is wide open from the earlier explosives."

"Sergeant Johnson," said Lee over coms.

"Yes, ma'am," came the reply.

"Lieutenant Wasp is with me. We are preparing to enter the port-side docking bay. You and Jones enter from starboard. We'll work through and meet up at the bridge."

"Roger, ma'am," said Johnson. She paused, and then said, "Lieutenant Wasp, sir, how are the rest of the team members?"

"They are recovering and are in good hands with Dr. Zakany," said Wasp.

"Thank you, sir."

Then the green light came on, the airlock hatch opened, and Lee and Wasp jumped into the void.

CHAPTER 91

INTO THE VOID

Lee thrust out from the airlock into space. She felt again the thrill and, if she were honest, the terror of free fall in space. She recalled the first time she had done this. It was when she was with the Marine Commando unit leaping from an asteroid toward the moon base Alpha 51.

On that mission she had gone into a spin and had to adjust her stabilizers to bring herself under control. It was a wild and dangerous ride. Upon landing on the Moon, she had come in hard and broken her jaw, three ribs and received a concussion that caused her to lose consciousness for a few minutes. The Marine next to her had been killed on impact.

Now Lee flew forward with Wasp at her side. As they approached the open hole in the side of the vessel, it looked to Lee like a gaping dark mouth with jagged teeth where the explosion had blown though the hull.

They both raised their weapons as they approached. Then they were through and into the vessel. It was completely dark, and Lee's system automatically switched to infrared. The hallway they were in was empty. The infrared picked up warmer spots that Lee assumed were residuals from the blast that opened the hatch.

"Johnson," said Lee into her coms. "We're in the port side. No contact yet."

"Ma'am, we're in starboard. No contact yet," reported Johnson.

"Raymond," asked Lee. "Any progress on the new batch of nanites?"

"Yes. Sending them out now," said Raymond. "They are programmed to follow your progress, give us early warning of any movement, and disable automatic defenses."

"Thank you," said Lee. "Let us know what they send."

"Small caveat," said Raymond. "They can't handle anything new. That's just a limitation of the technology. I've programmed them to handle anything that's already in our knowledge base. If it's something new, they won't be able to deal with it."

"That will have to do," said Lee.

Lee and Wasp moved cautiously down the passageway until they came to another hatch.

"Chief, what am I looking at?" asked Lee.

"That's the airlock," said Raymond. "It's still functioning; most of the vessel still has air pressure. I've set it so it will accept you to go through."

Lee and Wasp quickly negotiated the airlock and stepped into a new passageway, this time with light and air.

"We're through the airlock on the starboard side," said Johnson. "No sign of resistance yet."

"Acknowledged," replied Lee.

"Captain Lee, you have incoming from the corridor on your left," said Raymond.

"Let us see it," said Lee.

The image popped up on her heads-up visual display on her face plate. Two individuals moving quickly toward them from the left corridor. But something wasn't right; the two appeared to move awkwardly.

"They're not attacking; they're fleeing," said Lee. "They're being chased; see the two in battle armor behind them. Don't fire and let me handle it."

"All yours," said Wasp.

CHAPTER 92

A NAME FROM THE PAST

Wasp watched in amazement as the events unfolded in front of his eyes. Two large animals charged around the corner, slipping on the metal surface of the floor as they tried to get leverage, and headed directly toward them. He resisted the urge to raise his weapon and fire.

The larger of the two animals was covered in brownish fur mixed with lighter colors. The smaller one was almost all white, with only a smattering of darker shades on the underbelly. They ran on all fours. Their faces looked like that of a fox or a wolf, but they were clearly something else.

When the animals saw Lee and Wasp, they tried to stop, skidding for a few feet on the shiny surface. And when they did come to a stop, they showed their teeth, snarled, and raised up their spines in the classic posture of a cat preparing for a fight.

Both animals immediately looked back in the direction from which they had come.

The Lee did something that astonished Wasp. She pulled off her helmet, showing her face and jet-black hair spilling out. In a loud clear voice she said, "Ganzorig, I am Daichi Tengri. Run past us and we will protect you."

Lee gestured to show they should move past her and Wasp.

The animals' eyes went wide for an instant, and then they shot past Lee and Wasp, taking up a position facing the way they had come, but now behind the two Marines.

Mere seconds later two figures emerged from the adjacent hallway. They were in full battle armor and moving as rapidly as the bulky armor would allow. They saw Lee and Wasp and began to raise their weapons. And then they were dead. Lee and Wasp shot them simultaneously. One round each through their face plates.

Lee said on coms, "Chief, anything else coming our way?"

"Nothing close," said Raymond. "My nanites can't be everywhere. I'll let you know if that changes."

Wasp watched Lee as she turned toward the two animals. She spoke to them in a language Wasp didn't understand and held out her hand. One of the animals stood up on its back legs, walked forward toward Lee, stopped two feet in front of her and said in a very raspy voice, "Daichi Tengri." Lee then put out her hand palm downward toward the animal. It took Lee's hand in its paw and sniffed it.

Lee turned to Wasp and said, "These are Olgoi, the same intelligent species of diggers that helped us fight off the Nephilim."

"What are they doing here?" asked Wasp. "And how did you know his name?" He looked at her oddly, cocked his head and said, "Daichi Tengri?"

"Long story," said Lee. "Right now, we still need to take this vessel. I'm hoping Ganzorig can help us."

CHAPTER 93

NEW PLAYERS, NEW PROBLEMS

"Raymond, Danner, Laura" said Lee into her coms, "I know what the purpose of this vessel is. But first we are going to need to take it."

"What do you need?" asked Danner.

"Laura, how many of the Oracle team can we get back into the fight?" asked Lee.

"All of them," she said. "They are awake now and not happy about being held in sickbay. If I try to keep them any longer, they'll mutiny."

Lee turned to Wasp and said, "We'll need as many as can be made available to fight."

Wasp nodded and said into the coms, "Doctor, can you put Sergeant Wallace on coms?"

A moment later: "Wallace here, boss."

"You are in command of the team in sickbay," he said. "I need you, with all haste, to suit up as many as are able, armed, and ready for operations. We've made an initial entry here with just Captain Lee,

Johnson, Jones and me. The port and starboard docking bays are secure. Give me your ETA as soon as you're ready to go."

"Roger, sir, Wallace out."

Wasp turned to Lee and said, "What language are you using with the Olgoi? I have a translator program in my coms unit, but it isn't recognizing the language."

"It's a dialect of Mongolian. Try that and maybe the translator will get some of it."

Lee turned to the two Olgoi and asked, "Speaker Ganzorig, how many of the Olgoi are being held captive on this vessel?"

"One hundred forty-seven, Khatun," said Ganzorig. Lee knew the title Khatun was used to address a queen or woman of nobility.

Lee turned and pointed to the still-smoking bodies of the two she and Wasp had killed. "And those? How many are aboard this vessel?"

"Give me a moment, said Ganzorig." Then he spoke quietly with his companion, they placed their muzzles close together as if they were sniffing each other. Then Ganzorig turned back to Lee.

"By their scent, we know of 27 distinct individuals of their kind on this vessel. There could be more that we haven't smelled. And there are many more on the land outside the vessel."

The land outside the vessel? thought Lee.

"Can you show me where they are?"

Ganzorig shook his head and said, "It is difficult; you have no proper sense of smell. It is hard to explain."

Lee pulled out a hand-held device and held it up to Ganzorig. "This is a schematic of what we know about this vessel. The four green dots are where we are now. Can you show me on the diagram where your people are?"

Ganzorig sniffed the device and shook his head dismissively. Then he consulted again with the smaller Olgoi who also tried a sniff of the device, with a similar unsatisfactory result. Lee realized they were up against a true language barrier. The Olgoi probably communicated spatial relations via scents they issued.

Ganzorig said, "There is no useful scent on this thing."

"Speaker, it's not in the scent, it is a diagram on the screen," said Lee. "Please look at the screen."

Ganzorig looked at the screen, a clear look of consternation on his face. He stared, cocked his head to the side and then his eyes opened wide. He began chattering and sniffing the other, who also came close and looked at the screen.

Ganzorig looked back at Lee and said, "Yes, yes! We get it now. It is a child's drawing. I suppose if you can't smell properly, this is what you must use."

Lee said, "Please show me where your people are being kept."

Ganzorig used one of his pointed claws to tap the map.

"Got it," said Lee. "And the best route to avoid the enemy?"

Again, Ganzorig and his colleague conferenced and then he returned to Lee and traced out a route on the diagram.

"Enemy is here," pointing to an intersection on the map, "and here, and here."

"Ganzorig," said Lee. "You said there are many more 'on the land outside the vessel.' What did you mean by that? Are they in another vessel?"

"No!" Said Ganzorig with the first show of real frustration. "Outside the vessel! Not in another one. They have a camp on the land."

Lee shrugged and said, "This vessel is in space, very far from any land."

"Preposterous!" said Ganzorig. "Our people are forced to work in the mines, taken from this prison to the tunnels every day. I have been there many times."

"OK, then," said Lee.

She turned to her coms channel. "Danner, Raymond, did you monitor that conversation?"

"Raymond here. I have plotted the route to the captives' detention cell, and I've directed the nanites to try and scout ahead and secure the route for you. You will have to use the air ducts. It'll be tight."

"Roger," said Lee. "Sergeant Wallace, how are the preparations coming along?"

"Ma'am, we are in route, five minutes out. I've split the team to enter both port and starboard airlocks. I've contacted Sergeant Johnson and Corporal Jones. I've told them to stand fast until the team on their side arrives."

"OK, then, we have a few minutes to wait until the rest of the team arrives. Dr. Raymond, what did you make of Ganzorig's assertion that this vessel is on land?"

"It's a mystery to me," said Raymond. "Clearly the captive Olgoi are being transported to their planet. In their view this vessel is on the planet. But we've had this vessel under observation for some time and it clearly has not transited to another timeline."

"Could they be using another variation on your technology?" asked Lee. "You said earlier that they were using an excessive amount of power."

"I think that is a given, though I cannot see what it is they are doing," replied Raymond.

"Danner, do you have any thoughts about this?" asked Lee.

"Yes, I do," said Danner. "But we have another problem. A tactical force of a frigate and two corvettes has just entered the system. I don't know who they are, but they're definitely not Alliance vessels."

"Ok, Lieutenant Danner," said Lee. "Your job is to buy us time to seize this vessel. If possible, preserve the three shuttles under your command. If you need to jump away to do that, then do what you must. We are going to stay with this mission until we retake the vessel and free the Olgoi being held captive."

"Captain Lee," said Wasp. "The rest of Oracle are onboard now."

They turned to see Sergeant Wallace in full battle armor moving toward them from the direction Lee and Wasp had come. Behind her were five Marines also in full armor.

Lee turned to Wasp and asked, "Are you up to taking back command of your teams?"

"Yes, ma'am," said Wasp.

"Good, because we are going to split up," said Lee. "Use whatever combat power you have to take control of this vessel. We may need to jump away if the separatist warships now entering the system get too close."

"And what will you do, ma'am?" asked Wasp.

Lee looked at the Olgoi and said, "We are going to find the other Olgoi and set them free. And maybe they can help us out a bit."

CHAPTER 94

ADMIRAL CHAMBERS RETURNS

"Sir, you have an encrypted message on our Alliance coms," said Nemeth.

"Put it through to my screen," replied Chambers.

Chambers turned toward the screen at his desk and hit a key.

"Chambers here," he said.

On the screen appeared the image of a middle-aged woman, grey hair drawn back severely. She was wearing a flight suit, which showed the three stars of a vice admiral on her collar.

"Jay, Stephanie Bluefield here," said the woman.

"Admiral, it is good to see you," said Chambers. "How can I help you?"

"Jay, you've got trouble coming your way," she said.

Chambers smiled and said, "I recall you saying that once before on the Orion mission. Is it worse than that?" The Orion mission was a covert mission Chambers had led decades earlier. Bluefield, then a Naval lieutenant, had been the intelligence officer. At a critical moment she had given him a warning that saved his life and those of his team.

"Yes, Jay," she responded, "It's worse than the local militia showing up at the wrong time."

"Best to tell me then," he said.

"First, I know what you're doing in the TOI-700 system," she said. "I also know you have been cooperating with the Oracle team now operating in the Oort Cloud."

"Understand," said Chambers, never one to say more than was necessary.

"We both have a problem and I need your help," she said, her face and voice indicating a grave concern.

"Tell me," said Chambers.

"The mining operation executed a fail-safe alert about six hours ago," she said. "We have detected a separatist base launching some sort of armed response. We believe that force — a frigate and two corvettes — is headed your way."

She paused and then said, "Can I assume you have people in harm's way out there?"

"They are actually your people, not mine," he said. "I'm just trying to run a criminal empire out here." He gave one of his rare smiles.

"Jay, I'm sending Alliance help as fast as I can. You know I'm not a fleet commanding officer. I'm scrounging for resources. I've got three cutters under my authority; we use them for courier duty. But they are fast and already moving. They don't have much firepower, but it's better than nothing."

"Anything else?" asked Chambers.

"Yes, I reached out to the young commanding officer of an Alliance cruiser, one Commander Jay Chambers, Jr.," she said.

Chambers said nothing.

"He broke away from a training exercise," she said. "Technically, he's deploying to TOI-700 without authority; he's AWOL. I sent him a flash traffic war movement order, copied it to his fleet commander. I don't really have the authority to do that, but it should cover him, at least for now."

"I am grateful," said Chambers. "My son would not do something unless he felt it was the right thing to do. What cruiser is he commanding?"

"It's the Ticonderoga," she said. "It's a guided-missile cruiser."

"What do you need from me?" asked Chambers.

"I need time, Jay," she said. "These assets I'm sending are not going to be enough. I need you to delay the separatists, protect the Oracle force as long as you can."

Chambers turned toward Nemeth who had been on his communicator since the call came in from Bluefield.

Nemeth said, "Your shuttle is ready now. The Perseverance is powering up its warp drive in orbit. If we leave now, we can launch from orbit in 15 minutes."

Chambers turned back to the screen, "I've got an old D-Class frigate, we named it the Perseverance. It won't hold up for long against a modern frigate. I'll do what I can."

Chambers watched the screen as Admiral Bluefield put on her reading glasses, pulled out a document from an envelope and began reading.

"Jay W. Chambers, by the order of the Assistant Chief of Naval Operations for Personnel, you are hereby recalled to active duty at your retirement rank of Rear Admiral. You will take command of..." Here she paused and wrote something on the sheet. "The Alliance frigate Perseverance and proceed with all haste to protect Alliance personnel in your area of operations."

She put down the paper, removed her glasses and asked, "Any questions, Admiral?"

He smiled and said, "Is that real? Or did you just make it up?"

She gave a mischievous smile and said, "It will be real. The personnel guy owes me big time."

"I need my chief pilot to be restored to duty status as well," he said. "I trust you to do the paperwork."

She smiled and said, "Should that be for Chief Nemeth or Chief Attila?"

Chambers smiled at the joke. It was her way of saying she knew enough about his operation to know what aliases they were using.

"Please make sure the commanding officers of the cutters understand they will be under my command, same with the cruiser. I don't want them to go in piecemeal against a frigate."

"Already done," she said. "I would appreciate getting cutters back intact. They are the only vessels I have under my authority."

"No worries," said Chambers. "I will use the cutters for recon only. You'll get them back. How much time do I have?"

"You've got about two hours before the bad guys show up in system."

"I'm leaving now," said Chambers. "Thank you, Admiral."

"Thank you, Admiral," she said with emphasis.

CHAPTER 95

THE GREAT ESCAPE

Lee, Ganzorig, and Batbayar crept along an air duct that ran parallel to the hallway that led to where the other Olgoi were being detained. She let Ganzorig lead the way, as he had explained this route is the way they had used to escape earlier.

It turned out the other Olgoi, Batbayar, was Ganzorig's daughter. Ganzorig explained that his daughter, although a juvenile, was considered to be a competent warrior, and was being groomed to be the commander of the pack's military arm.

"We are over it now, can you smell it?" asked Ganzorig.

"No," said Lee. "My sense of smell is not as good as the Olgoi."

"Raymond, can you see where we are?" she said into her coms.

"I got you," he replied. "You are over a storage room. I can see the heat signatures of the Olgoi below you, but not much more. Most of the nanites are with the two Oracle teams."

"Can you show me where they are?" she asked.

"Doing that now," said Raymond.

On her handheld display, Lee saw a schematic of the vessel. She was several levels above the two teams, which were moving parallel to each other toward the bow of the vessel where the bridge was located. Because of the earlier electric shock that blew the majority of the team away, Wasp had ordered the teams to proceed with caution. They cleared every adjacent passageway and room looking for booby traps before proceeding forward.

Lee saw something on the display that disturbed her. "Chief, why can't I see what's forward of the bulkhead leading into navigation?"

"It's a bit of a mystery," said Raymond. "Of the nanites I sent past that point, none have returned or communicated back to me. I've stopped sending them that far; I can't afford to lose any more."

"Got it," she said.

Turning to Ganzorig, she asked, "Speaker, can you tell me what is forward of this point?" She handed the handheld device forward to him.

"Yes, many bad ones there," he responded. "And also, the door leading to the mines and their camp."

To Ganzorig, she asked, "How can we get down to where your people are, and how do we get them out?"

"Watch," he said. He turned to his daughter, and they nuzzled and chattered briefly. Then she squeezed past Lee and, using her claws and her teeth, pulled up some of the shielding that was the bottom part of the duct. It took a few tries, but soon there was a gap about two feet across.

Batbayar scooted backward giving Lee enough room to shuffle forward and look through the opening. She saw dozens of glowing yellow eyes looking up at her.

"My God, they are packed in together," said Lee.

"Yes," said Ganzorig. "It is very harsh."

"Where is the exit?" asked Lee.

"On the far side," Ganzorig pointed with his paw in the direction Lee knew was toward the bow of the vessel. "It is locked until they come for us or bring us back."

Just then Wasp broke into the coms. "We've got contact, both teams."

Lee could hear explosions and gunfire over the line.

"Status?" said Lee.

"We got hit with anti-personnel mines," said Wasp. "Then the bulkhead doors behind the teams have shut. Now we have fire from the front. The enemy has heavy armor suits, and we haven't been able to push through or withdraw."

"Understand," said Lee. "Chief, can you help them out?"

"Working on it," said Raymond.

"While you're doing that," said Lee. "I could use your help here."

CHAPTER 96

TASK FORCE ULYSSES

The Perseverance came out of jump 400 kilometers from the storage vessel. Chambers said, "Chief Nemeth, have the other Alliance vessels arrived yet?"

"One moment," said Nemeth. Then: "Yes, sir. The Ticonderoga and the three cutters are on station, hidden well within the Oort Cloud. Lieutenant Danner has three civilian shuttles, all armed, which he says are at your disposal."

"Please put them all on screen for a conference call," said Chambers.

Chambers saw the screen quickly populated with the images of the seven commanding officers of the various vessels now at his disposal. He paused a moment and looked at the faces of each of the captains in turn. Then he spoke:

"Captains, I am Rear Admiral Jay Chambers. I am the commanding officer of the Alliance Frigate Perseverance. As the senior officer present, I am primary for this task force, which will henceforth to be called Task Force Ulysses."

He paused to let that sink in. Chambers learned decades ago the best way to exert authority is to do so with complete confidence, and to never rush his speech. No one responded.

"Second in command is Commander Chambers of the Alliance Cruiser Ticonderoga." Again, he paused to look at his son on the screen. The last time he had seen him they had been adversaries in battle. That had been two years ago.

The younger Chambers had a mix of Asian and Caucasian features. As always, his expression showed nothing. It was the cold face, the same expression Lee used when going into battle. The same that their ancestors, the Mongols, had used when fighting under the great Khans.

Chambers continued. "Our mission is to protect the Alliance personnel on the mining storage vessel, Octavia. Coordinates are being sent to you now. That vessel is a key asset. There is an Alliance special operations team aboard now, whose mission is to take control of the vessel. We have been alerted by Admiral Bluefield of Naval Intelligence that the separatists will send a relief task force consisting of a frigate and two corvettes. My intent is to confront the commander of the frigate, attempt to gain his surrender. Failing that, to engage the three vessels in order to prevent them from interfering with the team we have aboard the Octavia."

Another long pause, then he said, "I am convinced that the separatists will send more assets in due time; that this initial response is simply what was available on short notice. Therefore, we need to prepare for a larger engagement."

Chambers paused, looked at the faces on screen and said, "Here is my plan. First, Chief Raymond, I'm going to need some of your nanites..."

After the briefing, Chambers asked the commander of the Ticonderoga to stand by.

The admiral said, "Son, it's good to see you. Thank you for coming to my aid. I hope you are OK."

The young Jay gave a rare smile and said, "Actually, I think I'm a fugitive about now. I left in a hurry and without permission. You've done some of that. Got any advice?"

"Admiral Bluefield tells me she thinks she can cover for you," said the admiral. "She sent an emergency war movement order and copied it to your fleet commander. That should help."

"All in a day's work," said the younger Chambers. "I would have come, with or without it. I'm all grown up and can take my medicine."

"Jay, I hope when this is all over, we can spend some time together. We have some catching up to do."

"I would like that, Dad," said the younger. "But I have a question."

"Shoot," said the admiral.

"Where did you get that rust bucket?"

CHAPTER 97

ADMIRAL, WHAT ARE YOUR ORDERS?

Commander Jason Evans of the Confederation Frigate Alvarado was excited. Two years ago, he had been a 45-year-old commander in the Alliance Navy. He had been passed over for promotion three times. The Navy had informed him that if passed over a fourth time, he would be required to leave the Navy. Worse, on his last operational assignment, he had been relieved from duty as the commanding officer of a troop transport by some washed-up rear admiral who had come on to his bridge after duty hours and presumed to take over his vessel. When Evans had tried to reestablish a sense of order over his incompetent bridge crew, the admiral had him arrested and confined to his quarters.

All because he had made a few mistakes. He didn't understand it. People make mistakes. Other officers had made mistakes and they were promoted. Just because he had shown up once a little hung over. Well, maybe more than once. Who didn't sometimes need a little something after a stressful day dealing with idiots who couldn't follow orders? He could function better while drunk than most officers fully sober. Why didn't they get that? At least now, as captain of a separatist warship, no one bothered him about that. In two years, he hadn't actually met in person any of his superiors. He liked it that way. If he wanted a drink now and then, that was his business.

When the crisis hit the Alliance two years ago, he had been offered a promotion and a command if he would join the separatist Navy. He had jumped at the chance and now he was back on duty as a full commander and the CO of a frigate.

Right now, on this mission, he was put in charge of two other vessels, both Corvettes. Task force commander. He liked the sound of that. He had been ordered to go immediately to the Oort Cloud of the TOI-700 system and check out a certain transport vessel that had failed to send in an all-clear report for several hours. He wasn't told why the vessel was important, or why three warships were required to respond to an overdue fail safe. It was probably nothing, just a communications failure.

"Sir, we are dropping out of FTL. Our scans show an older frigate blocking our path to the transport vessel," said the duty officer.

"Affiliation?" asked Evans.

"Our records show it listed as an orbital defense vessel from the planet Ulysses," said the duty officer. "However, the IFF response now shows it as an Alliance vessel, the Perseverance."

"Armament?" asked Evans.

"Not much, sir," said the officer. "Our scan can detect four missiles, all older Mark 74s. She has two five-inch guns. It is possible there's more; she has some modifications I don't recognize."

"And it's alone? No other Alliance vessels in the area?"

"We don't see any on scan, sir," said the officer. "Nothing's comes back on IFF. But we are in the Oort Cloud; plenty of places to hide."

Evans was troubled. He didn't like ambiguity. If he was honest with himself, he didn't like taking responsibility for difficult decisions. He always arranged his command so that someone else would take the blame if things went wrong. Why was an old rust bucket blocking his way? Since when did the Alliance use older vessels as warships? This didn't make any sense.

"Sir, the corvettes are asking for guidance," said the duty officer.

"Damnit! Tell them to standby, I haven't had a chance to evaluate the situation."

"Sir, the Perseverance is hailing you."

"What the hell am I supposed to tell him?" shouted Evans.

The bridge was silent. They had seen their captain go into a rage before when something — anything – happened that he didn't want to deal with.

"Put it on screen, then," said Evans petulantly.

On screen appeared the image of an old man with a wicked scar down the left side of his face. Evans had seen it before.

"You!" Evans shouted, sputtering with real anger. Chambers was the bastard that had relieved him of command of the troop carrier. The memory caused him to burn with humiliation.

"Weapons officer!" shouted Evans. "Prepare to fire on that frigate."

"Sir," said the Petty Officer of the weapons station. "I am not getting a response from the weapons console."

"Goddamn it," screamed Evans. "Do your job or I'll find someone who will."

The petty officer helplessly tapped keys but clearly nothing was happening.

Finally, Evans, now red-faced with frustration, looked up at the screen and said in a high-pitched, whiny voice, "What the hell do you want, Chambers?"

Chambers continued to look at Evans and remained silent.

"Well?!" screamed Evans.

In a steady, measured voice, Chambers said, "Commander Jason Evans, I have a standing order to arrest you on the charge of treason against the Alliance. You will consider yourself under arrest and turn your command over to the next senior officer."

"Not this time, Chambers," said Evans in a triumphant manner. "I have you outgunned and overmatched. It is you who will be relieved."

"Are you sure?" said Chambers in his eerily calm voice.

"Captain," said the voice of one of the bridge crew. "Radar shows the Alliance cruiser, Ticonderoga, just coming out of warp directly aft of us. It has a missile lock on us and the two corvettes."

"Sir," said another voice from the bridge crew. "The corvettes are demanding instructions."

"What is wrong with our weapons?" demanded Evans. "I want to fire on that damn frigate!"

"Sir, I don't know. I'm not getting any response."

"Commander Evans," said Chambers in that same calm voice that infuriated Evans. "No need to harangue your crew. I have control of your weapons, as well as the rest of your combat systems, including your communications. And now I need to speak to your crew. Ship-wide channel, please."

Evans heard the three-tone alert indicating someone was about to address the crew over the ship-wide audio system.

"Shut that down!" screamed Evans.

"I'm sorry, sir; nothing is responding," came the reply from the communications officer.

Chambers began to speak, and Evans realized with horror that Chambers was speaking to the entire crew of the Alvarado.

"Crew of the Alvarado, this is Rear Admiral Jay Chambers of the Alliance Navy. Your commanding officer, Commander Jason Evans, is wanted for treason against the Alliance. Up until this moment, each of you has been following what you believed to be the lawful orders of your commanding officer. I am authorized to grant amnesty to any and all of you at this time provided you cease taking orders from Jason Evans and accept the lawful authority of the Alliance. This offer of amnesty will remain open for ten seconds."

Chambers stopped, folded his arms, and waited patiently.

Evans looked around the bridge wildly and saw only hard stares from the crew. "Damn it," he said. "Prepare my escape pod, I'm getting out of here. You disloyal bastards can take your chances. Don't think for a second, he'll honor that amnesty offer." When no one moved, he screamed, "Well? Get moving, somebody!"

Then he felt cold steel at the nape of his neck. He froze unable to speak for terror.

"Sir, this is Lieutenant Commander Bartos, the executive officer," said the voice behind him. "You are hereby relieved of command."

"Bullshit!" shouted Evans. "If you think I'm going to hang while you pricks get a free ride..." Evans stood up, whirled around, and grabbed awkwardly for the pistol in Bartos' hand.

The shot rang out loudly in the enclosed space of the bridge. Evan's head jerked back as the bullet entered his forehead and exited the back of the skull causing a spray of blood and grey matter.

"Provost, please remove these remains," said Bartos to the Petty Officer in charge of security on the bridge.

Once that task was completed, Lieutenant Commander Bartos sat in the captain's chair, faced Chambers, and said simply, "Admiral, what are your orders?"

CHAPTER 98

NO FEAR

Corporal Baxter Jones felt despair for the first time in his nine years in the Marine Corps. His team, led by Sergeant Wallace, was pinned down under heavy fire. Because of the narrow passageway, they had no cover, and they couldn't find a target to aim at. The enemy was using an automated gun system that would pop out of the walls or ceilings at random intervals, fire a burst and then withdraw. They hadn't been able to disable it. And his team couldn't withdraw because the corridor had sealed up behind them.

Sergeant Wallace had stopped responding over coms, so he could only assume she had been killed or wounded. That left him in command of the team, but he didn't know what to do. His armored suit had taken hit after painful hit, but it wasn't going to hold up forever.

After what seemed like an hour, he finally ran out of ammunition. Any surviving members of his team had also run out because there was no fire from his side. He stood up and drew his Ka-Bar knife. He knew he would die. But he also knew a Marine didn't go down without fighting.

He saw that two other Marines of his team also stood; both drew their knives. Then a third figure stood. It was Sergeant Wallace. She was barely standing. Her armor showed a dozen impact points, including one to her face shield, which was half shot away. He could see her bloodied face through the jagged opening. On one side of her face, her cheek hung loose, showing her bloodied teeth.

She looked at Jones, tapped her helmet's earpiece to indicate her coms were out, and gave a wicked smile. Then she drew her Ka-Bar and nodded. Jones thought he had never seen anything so beautiful. No one said a word; there was nothing to say. They were Marines and they knew what to do.

At the end of the corridor, Jones saw the figures of half a dozen fully armored enemies walk toward them, weapons pointed directly at the Marines.

"Marines, prepare to charge," he said.

Just then something changed. The enemy stopped moving forward. They looked around as if confused. In synchronization, they each reached up toward their helmets and rapidly removed them. Jones could see distress in their expressions.

Then he heard a sound, not like anything he had ever heard before. It sounded like a murmur of voices, eerily deep and guttural, rising in volume as it got closer. The enemy turned back to face whatever was making the sound. They raised their weapons, and then all hell broke loose.

Jones saw what looked like a white wolf leading a pack of brown wolves racing around the corner from behind the enemy. Teeth bared, howling in that unearthly sound, they leapt upon the enemy, tearing their exposed faces and throats. The screams of the enemy were horrible to hear, but thankfully cut short quickly.

The four remaining Marines stood shoulder-to-shoulder facing the animals, knives extended in a defensive stance. The white one walked forward slowly, blood on its jaws and the fur of its face and throat. In a deep, gruff voice, it said something Jones' headset translator interpreted as "No fear."

CHAPTER 99

OBJECTIVE SECURE

Lee, Wasp, Johnson, Ganzorig, and Batbayar stood on the bridge of the storage vessel, Octavia. They faced the viewscreen, which showed two images. On the left was open space, dark but for stars in the distance. On the right showed a view of a landscape in bright sunshine. A smaller view screen below showed split-screen images of Raymond, Danner, and Chambers.

Lee said, "OK, let's take stock. Admiral Chambers, can you let us know your status?"

"I am in command of an Alliance task force consisting of a frigate, a cruiser, and three cutters. The frigate is from Ulysses' orbital defense force, conscripted into Alliance service. Admiral Bluefield sent the cruiser and cutters to assist in the defense of the Oracle team aboard the Octavia. We confronted and forced the surrender of a separatist task force consisting of one frigate and two corvettes."

"Do you expect more separatist forces to be sent to the system?" asked Lee.

"Perhaps," said Chambers. "But Admiral Bluefield is in route, leading an Alliance task force sufficient to deter any move by separatist forces."

"Thank you, sir," said Lee. "At this end, the Octavia is secure. The Oracle team took heavy casualties: of the twelve committed to the operation, two are dead, four are seriously wounded." Lee paused for a moment, looked at Wasp before continuing. He was clearly struggling with the magnitude of the loss.

"Lieutenant Wasp, I am sorry for your losses," said Chambers.

"Thank you, sir," said Wasp. "The wounded are being well-cared for by Dr. Zakany and our corpsmen."

"Captain Lee, I understand you were able to free some captives," said Chambers.

"Yes, sir," replied Lee. "The separatists aboard this vessel were holding 127 of the Olgoi as forced labor. Their leader, Speaker Ganzorig, and his daughter Batbayar are with me now."

Lee turned to indicate the two Olgoi. "Batbayar led the group of Olgoi who defeated the separatist aboard this vessel. They speak a dialect of Mongolian. We are using an automated translator so they can speak and understand."

"Speaker Ganzorig, warrior Batbayar," said Chambers. "We are grateful for your valor that saved the lives of our Marines."

"And we are eternally grateful to Daichi Tengri and her valiant warriors who sacrificed so much to free us. But the battle is not over. Many are still held hostage on the planet."

Lee said, "I suppose that requires some explanation. Chief Raymond, can you shed some light on this?"

"Yes, Captain Lee," said Raymond. "I can explain the what, but not all of the how."

"Please do," said Lee.

"First, I am so very sorry my nanites couldn't disable the automated gun systems in the corridors," said Raymond. "I tried everything; I just didn't have enough of them. I finally hit on the idea to cut off the enemy's air flow in their armor. When they took off their helmets to breathe, the Olgoi attacked them. None of the enemy survived."

"What about this vessel?" asked Lee.

"I don't know how they are doing it," said Raymond. "But the Octavia is being held in two different timelines at the same time." He turned to the split screen. "What you are seeing is a simultaneous view of space as we see it, and the planet of the Olgoi. This vessel is in both places and times concurrently. This explains the requirement for the incredible power usage by the Octavia that we noted earlier."

"But why do this?" asked Chambers.

"The timeline for the planet is approximately 800 years in our past," said Raymond. "The separatists have been holding what we thought was a storage vessel in both timelines to allow the transport of the Boron Nitride directly to the planet. They captured and compelled the Olgoi to bury it for them."

"And their purpose for doing this?" asked Lee.

"So that it will be available in the present day for whatever time-travel purpose the separatists plan to use it for," said Raymond.

"So, what you're saying," said Lee. "Is that the planet today has large quantities of Boron Nitride buried in its crust."

"Yes," said Raymond. "The presence of that ore in the crust would explain the abnormalities we observed from orbit in the planet's magnetic field. It is part of the reason we had trouble finding you when we surveyed the planet looking for you."

Wasp turning to Lee said, "Ma'am, are you saying Ganzorig and his people are from the planet 800 years before our time?"

"Yes, that is correct," said Lee.

"So how did you recognize Ganzorig when you saw him?" asked Wasp.

Lee turned to Ganzorig and said, "Speaker, I didn't actually recognize you. But I had been told by the speaker who is current in my timeline, Arslan, that Daichi Tengri had saved Ganzorig from danger and had called him by name. I knew I had to protect the two of you when I saw you fleeing. If you stayed where you were, you would have been killed. So, I took a chance that you would respond to my command to take cover behind us."

"And are you this Daichi Tengri?" asked Wasp with a wry smile.

"Daichi Tengri is the War Goddess of Mongolian mythology. I am who I am; and apparently one of my personas is Daichi Tengri. Beyond that, I have no explanation," said Lee.

"We have an immediate problem to solve," said Raymond.

All eyes turned to his image on the screen. "The Octavia is being held in two timelines and two places simultaneously by the expenditure of a great deal of power. I do not have a full understanding of how this works. I do know that it is not sustainable. I don't know how this power is being generated or sustained. What I do know is that if the power drops below some level of sustainment, the vessel will not remain where it is now."

"What will happen?" asked Lee.

"I'm not sure," said Raymond. "It will remain in one of the timelines and not the other, or it will jump to another random timeline and location."

"Let's evacuate the vessel," said Lee.

She turned to Ganzorig and said, "Speaker, you and any of your remaining people should leave this vessel immediately."

"As you command, Khatun," replied Ganzorig. "We will do all that we can to free our people still held in captivity. I will tell them Daichi Tengri is here to fight for them."

Just then the light on the bridge flickered, then went out altogether. A moment later a dim glow of emergency backup lighting came on.

"The vessel is losing power," said Raymond. "You need to get everyone off now!"

"Let's move it, people," said Lee.

Ganzorig and Batbayar both looked at Lee, bowed deeply, turned and hurried away.

Five minutes later both Oracle teams were loaded onto the two shuttles docked on the port and starboard sides of the vessel.

"Captain Lee," said Wasp. "We are loaded and ready to go. Waiting on you. How far away are you?"

"I'm not coming," said Lee.

Wasp was silent for a moment. He knew Lee, and he understood what she was doing.

"You will stay with the Olgoi, then?" asked Wasp.

"Yes," said Lee. "They have hundreds that remain captive. I'll stay until they are secured."

"I could leave a team with you to help," said Wasp.

"Negative," said Lee. "Time for you to go. I'll do..."

The transmission stopped abruptly, and Wasp watched as the Octavia faded into nothingness.

CHAPTER 100

AFTERMATH

Rear Admiral Jay Chambers was in the Commanding Admiral's ready room on the Flagship R. E. Lee. He sat with a mug of coffee across from Admiral Stephanie Bluefield, now sporting four stars on her collar.

"Congratulations on your promotion to full admiral," said Chambers.

"I owe my thanks to you for handling the separatist task force," she said. "The intel we got from the crews of the Alvarado and the two corvettes was invaluable. Apparently, the Chief of Naval Operations thought I was the right person to handle the situation going forward; thus, the promotion and the command of a battle fleet."

Bluefield looked at Chambers, cocked her head and said, "What about you? I could use you as my deputy. There's lots of work ahead. It would mean another star for you."

Chambers was thoughtful for a long moment and then said, "I'm tempted, Admiral. I'd be honored to take that position, and happy to work for you. And I think I'd be good for that post."

"I hear a but coming," said Bluefield with a smile.

"You know that Captain Lee is my niece," said Chambers. "I never have, nor would I ever, show her favor while I was serving as an Alliance officer. In fact, I left her to the tender mercies of the Alliance naval hierarchy after the battle of Alpha 51, knowing full well she would be targeted, maybe even killed."

Bluefield was silent, waiting for Chambers to finish his thoughts.

"She disappeared with the Octavia, and I'd like to try to get her back," said Chambers.

"Do we know where the vessel went?" asked Bluefield.

"Raymond thinks it is on the planet with the Olgoi back 800 years ago," said Chambers.

"Can't he just jump there and see if she is there?" asked Bluefield.

"Apparently not," said Chambers. "Raymond tells me the timeline she went to is no longer closely aligned with ours. He can't see it, and therefore he can't jump to it."

"What are your options?" asked Bluefield.

Chambers sighed and shrugged his shoulders in a manner Bluefield felt was uncharacteristic of the Jay Chambers she had known for decades.

"I can't think of anything, and Raymond says he can't help," said Chambers. "He thinks she is lost and cannot reach us."

They were both silent for several minutes. Then Bluefield said, "Jay, it's not like you to give up. May I suggest a course of action?"

"Yes, of course," said Chambers.

"Take the post as my deputy. It's not an offer of charity. Except for my short time in special operations with you 20 years ago, I've been in intelligence my whole career. Now I'm commanding a battle fleet of dozens of combat vessels. I will need a deputy with more experience in operations. You have that experience and more importantly, I trust you."

Chambers shrugged noncommittally.

"Take the job," she said. "We have work to do, and I promise you, if we can recover Captain Lee, we will. I won't do it as a favor to you or to her. She is an Alliance officer missing in action. We do have a duty to every one of our people missing to make every effort to recover them if possible."

Chambers looked up at Bluefield and said simply, "Ma'am, I accept your offer."

"I'm relieved, Jay. I need you here. I'll have your things transferred to the command vessel."

"I have another reason for taking the post," he said. "You should know it before you agree to take me."

Her face turned to a serious expression, and she said, "Tell me."

"Raymond says there is no risk of us fiddling with the past of another timeline, that we can't create a paradox."

"That's what I understood, as well," she said. "We can't change our own past because we aren't actually traveling to our timeline. What about it?"

"He's wrong," said Chambers.

"Why do you say that?" she asked. "Raymond is probably the greatest physicist of this century, maybe any century. Are you sure he is wrong?"

"Yes," replied Chambers simply. "He has to be. When he goes back into another timeline, some other version of Raymond comes into ours. Because they are nearly identical, whatever he does to their timeline, some doppelgänger Raymond does to ours. And that changes our timeline, our past, our present and our future."

"How does this affect your decision to accept the post as deputy commander?" she asked.

"Time travel is the fulcrum for anything that happens now," said Chambers with an uncharacteristic forcefulness. "Whether it's the separatists or our own people, we need to be careful. We need to stop it if we can. And this task force is best placed to make those contributions. I do want to be part of it."

Admiral Bluefield stood, a grim expression on her face. She extended a hand and said, "Welcome aboard, Vice Admiral Chambers."

EPILOGUE

Laura Zakany was excited, happy, filled with an anticipation that left a tight knot in her stomach. She had caught a ride on a frigate heading to Earth. From Earth orbit, she caught a shuttle. Now she sat in a comfortable window seat of that shuttle which was bringing her back to Perth. Her trip had taken two weeks, and the anticipation of seeing Matthias, Jr. again had built up that entire time.

She was finally going home to see her son, the love of her life. Looking back, she understood now what it meant to love one's own child. That was something you didn't come to understand without experiencing it. The love for your child was all encompassing, it dwarfed every other emotion.

But Laura had left her child — been away for months — to go on this insane mission in hopes of finding her husband. She had done what had been asked of her. Lieutenant Laura Zakany had fought pirates and monsters. And she had done her duty as a physician.

Her husband Matthias was back in her life, back in their lives. She had told him he could stay on with his admiral to continue the fight. They all knew the stakes were high now that the separatists had discovered time travel. But they had agreed he was never to disappear again, no matter the operational necessity.

But now, finally, she was going back to her primary duty: to be a mother to her son, Matthias, Jr., now two-and-half years old. She knew she would never leave him again, not like this on some crazy mission light years away with no known endpoint. Not for her husband, not for anyone or anything, would she ever leave her child again.

The trip back to Earth had been unsettling. She had access to news feeds, and there was trouble on her home planet. Some sort of mass hysteria had broken out. People were rioting, fighting, delusional. And it didn't seem limited to any specific location. Reports from every country said the same thing: people were frantic, some claiming strange, even bizarre things.

She was not aware of any medical condition that could cause such schizophrenic symptoms to such a significant portion of the population across geographical boundaries. The symptoms had appeared simultaneously everywhere on Earth starting about a week ago. This argued against some sort of pathogen. If it was a virus, it would have had a known origin point and time, and then spread in predicable ways depending on whether it was airborne or transmitted via some other way.

As the shuttle approached the coastline of Australia from the West, she looked out the window and noticed something was different. First, the coastline was several kilometers further west than it had been when she had left four months ago. She wondered what type of ecological event could have caused such a change. She thought, how does the Indian Ocean pull back thousands of meters in four months?

Next, she saw that parts of the shoreline of the city of Fremantle were reduced to blackened ash, as if there had been a great fire in the western part of the city. The Dampier Peninsula, cut out from the mainland by the Swan River, was completely black with not a single structure still standing.

Her heart raced as she realized some catastrophe had taken place within a few miles of her son while she was gone. But why hadn't she been informed? A recent fire of that magnitude should be all over the news.

Her aunt Anna had moved into her quarters on the base to take care of Matthias while she was gone. She quickly used her civilian com link to reach Anna, but the system came back with a recording that there was no such number.

She immediately contacted the provost marshal for the base.

"Provost Marshal's office, Petty Officer Franklin speaking. How may I help you?"

"This is Lieutenant Zakany. I need a welfare check on my address. My two-year-old son and my aunt are staying there, and I can't raise them on coms."

The Petty Officer took down the address and said he would dispatch a patrol car to that location right away.

"Ma'am, if you want, you can stay on the line. A patrol car is in the area and should be there in a few moments."

"Thank you," said Zakany. "I'll stay on the line. I've been gone four months. So, what happened on the west coast? I'm in a shuttle coming in and I saw the fire damage on the peninsula. And what caused the water level to drop?"

There was a long silence that stretched out uncomfortably. She could hear a muffled discussion in the background, as if the petty officer had covered the microphone and was speaking to someone near him. She thought she heard, "I have another one, a lieutenant inbound on a shuttle." A pause, then a response she could not make out.

"Lieutenant," said a new voice on the line. "I'm commander Benton Frazier, the provost marshal for the base. I'm sending a car for you to the spaceport. It will meet you at the shuttle."

"Commander, what's going on?" Zakany asked in desperation. "Is my son OK?"

"Better that I explain to you in person; wait." There was a pause and some muffled conversation. She couldn't understand most of it. But she heard Frazier say, "Hold it for me."

When he came back on the line it was clear from the background sounds that he was moving as he spoke. "Lieutenant Zakany, I am on my way to you now. Do not depart the spaceport by any other means. I will clear you through customs."

"What the hell..." she started to respond, but the line went dead.

When the shuttle pulled up to the ramp at the spaceport, Laura could see two military police vehicles on the apron, both with flashing emergency lights. When the shuttle door opened, a military police sergeant entered and said, "I need everyone to remain seated." He turned to Zakany and said, "Lieutenant, you're with me." Not waiting, he turned and departed.

Laura quickly grabbed her carry-on bag and followed him out and down the stairs to the waiting vehicles. The sergeant was standing by the open door, clearly waiting for her to enter the car.

Once inside, she saw sitting opposite her a young naval commander. He had on his left arm the brassard with the reflective lettering, MP. As the car pulled away, he looked at her with a sad, searching look, and Laura feared the worst.

"Please just tell me," she said through a stifled sob. "Is my son alive?"

"Lieutenant," he said slowly. "Laura, I am going to tell you everything I know. I promise I won't lie to you, and I promise I won't hold anything back."

Laura nodded, unable to speak.

"We have no record of your son, Matthias."

"What?" said Zakany unbelievably. "I gave birth to him here, at the base hospital. I left him here with my aunt Anna Zakany four months ago."

The commander nodded as if he understood. "You are perhaps the 50th person to report to the PMO that a loved one or friend is missing.

In most cases, we have no record of the missing person. In some cases, that person has been deceased for some time."

"What about my quarters?" she asked. "Has your patrol checked it out?"

"Yes," he said in a steady voice. "The address you gave for your quarters does not exist."

"But I lived there. I raised my son there."

"I'm positive," he responded. "Just to be sure, I had the driver take me past that address on the way here. It is an empty lot and has been for 50 years. Our records show you have resided at the bachelor officers' quarters since you graduated from medical school. I sent a team there to check it out. It is definitely your apartment. Your things are there."

Zakany sat in stunned silence. She knew this wasn't right. The commander must have made some huge mistake.

Then she thought of something. "Does this have something to do with the pandemic?" she said. "All these people with schizophrenic symptoms? Are you feeling OK, commander?"

"Believe me, Lieutenant, I'm fine."

"You don't seriously think I'm infected?" said Laura. "I'm the doctor, remember. I'd know if I had symptoms or hallucinations."

The commander said nothing.

She paused, cocked her head, and said, "Wait, I'll prove it to you. I have photos of my son." She reached into her breast pocket and retrieved her handheld device. She tabbed through the photo app and froze.

"There are no photos of your son on your handheld," said Frazier. It was more of a statement than a question.

"How did you know?" said Laura in near panic. "Did you wipe my phone?"

"No, I didn't," he responded. "Every person that reported a missing loved one did the same thing. They went to their photo app and couldn't find any personal photos of the missing person. Not on their app, not on the web, not in hard copy, not anywhere."

Frazier looked pained and Zakany said, "There's something else, isn't there?"

Frazier nodded and said, "It's not just missing people, it's added people."

"What does that mean?" asked Laura sharply.

"People are showing up at their homes and claiming they don't recognize their spouses. They are adamant that they are not married to the person they have been living with for years, sometimes decades."

"And children?" asked Zakany in a trembling voice.

"Yes," he replied. "We have parents who don't recognize the children and vice versa."

Zakany looked out the window of the car and asked with alarm, "Where are we going?"

"There is an operations center set up in the basement of the old base library," said the commander. "You have been requested by name to serve on a task force set up to deal with the crisis."

Zakany was quiet for a moment and then asked, "Do you think I'm sick? That I'm infected?"

"I don't know what's happened; but no, I don't think you are sick. And neither does the psychiatric team that is already working on this. What I've been told is that the people affected by this... whatever it is... have only one symptom."

"And that is?" asked Zakany.

"They insist there is a relationship with a person who is not present, or that there is no relationship with someone that is here."

"Your point?" asked Zakany.

"A true psychosis never presents that way. Instead, a psychotic person demonstrates a portfolio of symptoms and behaviors. Other than being frantic and losing a loved one, the vast majority of those impacted don't have any ancillary symptoms. No voices, other delusions, hallucinations, or manic behaviors."

"And what do you think?" she asked.

"I'm not an expert in psychology, but as a law enforcement officer, I am trained to identify abnormal behavior as a precaution in case a suspect may be dangerous. You do not have those behaviors, nor do most of those who reported to the PMO about a lost loved one. Something else is happening."

"What about the Dampier Peninsula?" asked Laura. "It looked like a war zone. Fremantle is in ruins. What happened?"

"That is another anomaly," said the commander. "Affected people don't recognize what should be familiar surroundings. The entire coastline was destroyed during the attack almost three years ago. Some of it has since been rebuilt, but most of it has been, and remains, a desert."

"That's not how I remember it," said Zakany with emphasis. "It was not hit; I actually took Matthias there to the beach for his first birthday; it was fine."

The commander looked straight forward, his jaw tight, and said nothing.

"What?" said Zakany. "You remember it differently?"

"Yes," he said. He turned to her with a somber look on his face. "My wife was on the peninsula when it got hit. I lost her. That's what I remember."

"And the coastline?" she said. "It is extended several kilometers further West."

"That also happened as a result of the attack," he said. "The dust in the atmosphere caused a global cooling. The ice caps at the poles have rapidly expanded, and sea levels worldwide are dropping."

Zakany shook her head. "Was it like that when I left four months ago?"

"Yes," he said simply.

Their vehicle pulled up to the old brick building that served now as an operations center. There were several police cars out front, all with their police flashers on. A uniformed MP opened her door and the two of them were hurriedly ushered down steps leading to a doorway guarded by two more MPs.

Once inside, they walked down a musky-smelling corridor, past another checkpoint and into a large, brightly lit room busy with uniformed people working on computers, digital displays and speaking earnestly into communicators.

They headed toward the back of the room, where Zakany could see the backs of several people facing a screen that seemed to be showing a map of the world. She recognized major population centers, lit up by indications in red, showing numerical figures of some sort.

In the center of the group was an older man in a Naval uniform. By the stripes on his sleeve, she could see he was a Vice Admiral. He had longish grey hair, and even from the back, she could see he had a full white beard.

Commander Frazier said, "Sir, as you requested, Dr. Zakany is here."

The man stopped what he was doing, stood straighter, but didn't turn to face them.

Finally, Zakany said, "What the hell is going on here?"

The man answered in a familiar voice, "Mei Ling has changed the past..."

He turned, and Laura Zakany gasped. She saw Admiral Jay Chambers. He looked different. Despite the full beard, she could see he had no scar on his left cheek and his left hand was unmistakably flesh and bone.

"And that has changed everything," he finished.

The End of Book Two

ABOUT THE AUTHOR

Following his retirement after 25 years as a US Army Paratrooper and Ranger, Brendan Wilson worked for 15 years at the North Atlantic Treaty Organization (NATO), where his portfolio focused on diplomacy and defense planning. He has served in war torn areas in Kosovo, Bosnia, Libya, Ukraine, Korea, and Iraq.

In addition to *The Achilles Battle Fleet:* Book One of the Mei Ling Lee series, he has served as a writer and executive producer for two award-winning short films. *Doug's Christmas* was an official selection at the 2014 GI Film Festival and won best picture at the Route 66 Film Festival. He wrote and produced the film, *A Child Lies Here* which had 12 official acceptances, eight nominations, and two awards including best actor and emerging filmmaker. Mr. Wilson served as the executive producer for the award-winning web series, *Greetings! from Prison*, which won Best Comedy Web Series at the LA Web Fest and the Idlewild International Festival of Cinema. He wrote, produced and acted for two martial arts documentary/training films made in Greece.

As a martial artist, he achieved grandmaster rankings, coached military martial arts competition teams, and developed a unique form of martial art called Aristos based on the principles of classical Greece. In 2009, Mr. Wilson won the silver medal at the US Open for Taekwondo. In 2022, he was inducted into the American Martial Arts Alliance (AMAA) Hall of Honor, and in 2024 he was recognized as Grandmaster of the Year by Action Martial Arts Magazine.

Mr. Wilson has a PhD from Berne University and a JD from Northwestern California University School of Law.

Brendan lives in DeKalb, Illinois, where he spends his time writing, practicing martial arts, and hiking with his beloved wife, Kay.

www.ingramcontent.com/pod-product-compliance
Lightning Source LLC
Chambersburg PA
CBHW021429240626
47153CB00001B/79